SpringerBriefs in Applied Sciences and Technology

# SpringerBriefs in Computational Intelligence

SpringerBriefs in Computational Intelligence are a series of slim high-quality publications encompassing the entire spectrum of Computational Intelligence. Featuring compact volumes of 50 to 125 pages (approximately 20,000-45,000 words), Briefs are shorter than a conventional book but longer than a journal article. Thus Briefs serve as timely, concise tools for students, researchers, and professionals.

Krishna Suryanarayan

# Formal Proofs in Elementary Set Theory

## Rules of Logic for Formal Proofs in Zermelo Set Theory

Springer

Krishna Suryanarayan
Department of Electrical Engineering
Indian Institute of Technology Madras
Chennai, Tamil Nadu, India

ISSN 2191-530X ISSN 2191-5318 (electronic)
SpringerBriefs in Applied Sciences and Technology
ISSN 2625-3704 ISSN 2625-3712 (electronic)
SpringerBriefs in Computational Intelligence
ISBN 978-3-032-22238-1 ISBN 978-3-032-22239-8 (eBook)
https://doi.org/10.1007/978-3-032-22239-8



This Springer imprint is published by the registered company Springer Nature Switzerland AG
The registered company address is: Gewerbestrasse 11, 6330 Cham, Switzerland

# Preface

A formal proof is a proof that is mechanically verifiable. The book gives the details of how to write a formal proof. The book gives formal proofs of theorems in set theory that is commonly employed as a foundation for the whole of mathematics. The Zermelo axiom system is chosen as the basis for set theory. The Zermelo axioms are sufficient for the topics covered in this book.

The first chapter of the book gives the rules of logic required to write formal proofs. It is assumed that the reader has basic knowledge of logic. A detailed justification of the rules of logic can be found in some of the references given at the end of the chapter.

The second chapter of the book gives formal proofs of theorems in elementary set theory. It includes a proof of the existence of a Peano system that provides a foundation for the number theory. The reader is expected to know what "axiomatic set theory" is.

The book helps the reader to write formal proofs. The book can serve as a reference for a course that covers "formal proofs" or "formal proofs in set theory." The book can be used to develop a software that can be used to verify formal proofs.

Chennai, India
December 2025

Krishna Suryanarayan

**Competing Interests** The author has no competing interests to declare that are relevant to the content of this manuscript.

# Contents

# Chapter 1
# Logic

This chapter gives the rules of logic that apply to stating axioms, stating theorems and developing formal proofs.

## 1.1 Symbols of Logic

The symbols of logic [1–4] are given in Table 1.1. In addition to these symbols, the punctuation marks or parentheses "(" and ")" are used. "=" is a predicate. The symbols of logic, variables, constants, predicates and operations form the set of building blocks.

## 1.2 Definitions

The following definitions are made in terms of symbols of logic, variables, constants, predicates and operations [1–5].

1. Term: Variables and constants are terms. The variables are denoted by $a$, $b$, $c$, ... or by $a_1$, $a_2$, $a_3$, .... The terms are denoted by $\alpha$, $\beta$, $\gamma$, ... or by $\alpha_1$, $\alpha_2$, $\alpha_3$, .... If $\alpha_1$, $\alpha_2$, ..., $\alpha_n$ are terms, where $n$ is a positive integer, then an operation with $n$ arguments applied to $\alpha_1$, $\alpha_2$, ..., $\alpha_n$ gives a term. For example, if $\alpha$ and $\beta$ are terms, then $\alpha \cup \beta$ is a term obtained by applying the operation "$\cup$" to two arguments $\alpha$ and $\beta$. In some cases, to avoid ambiguity, parentheses are used if an argument of an operation is obtained by applying an operation (not necessarily the same operation) to its arguments. For example, $\alpha \cup \beta$ is a term obtained by applying the operation "$\cup$" to two arguments $\alpha$ and $\beta$, and $(\alpha \cup \beta) \cup \gamma$ is a term obtained by applying the same operation to two arguments $\alpha \cup \beta$ and $\gamma$.



K. Suryanarayan, *Formal Proofs in Elementary Set Theory*, SpringerBriefs in Computational Intelligence, https://doi.org/10.1007/978-3-032-22239-8_1

**Table 1.1** Symbols of logic

| Symbol | Read as |
|---|---|
| $\neg$ | Not |
| $\vee$ | Or |
| $\wedge$ | And |
| $\Rightarrow$ | Implies |
| $\Leftrightarrow$ | If and only if |
| $\forall$ | For all |
| $\exists$ | For at least one |
| $\exists!$ | For exactly one |
| $=$ | Is equal to |

2. Atomic formula: If $\alpha_1, \alpha_2, \ldots, \alpha_n$ are terms, where $n$ is a positive integer, then a predicate with $n$ arguments applied to $\alpha_1, \alpha_2, \ldots, \alpha_n$ and enclosed within parentheses (called outermost parentheses) gives an atomic formula. For example, if $\alpha$ and $\beta$ are terms, then $(\alpha = \beta)$ is an atomic formula obtained by applying the predicate "=" to two arguments $\alpha$ and $\beta$ and enclosing within parentheses.
3. Formula: An atomic formula is a formula. The formulas are denoted by $A$, $B$, $C, \ldots$. If $A$ and $B$ are formulas, then $(\neg A)$, $(A \vee B)$, $(A \wedge B)$, $(A \Rightarrow B)$ and $(A \Leftrightarrow B)$ are formulas. If $a$ is a variable and $A$ is a formula, then $(\forall a)A$ is a formula if $A$ does not contain $(\forall a)$, $(\exists a)$ or $(\exists! a)$. If $a$ is a variable and $A$ is a formula, then $(\exists a)A$ is a formula if $A$ does not contain $(\forall a)$, $(\exists a)$ or $(\exists! a)$. If $a$ is a variable and $A$ is a formula, then $(\exists! a)A$ is a formula if $A$ does not contain $(\forall a)$, $(\exists a)$ or $(\exists! a)$. In the formulas $(\neg A)$, $(A \vee B)$, $(A \wedge B)$, $(A \Rightarrow B)$ and $(A \Leftrightarrow B)$, the parentheses enclosing $\neg A$, $A \vee B$, $A \wedge B$, $A \Rightarrow B$ and $A \Leftrightarrow B$ are called outermost parentheses. To simplify the appearance of formulas, the outermost parentheses are omitted; however, the omitted parentheses have to be added to the formulas if these formulas are used to construct certain other formulas. The rules for omission of parentheses and addition of omitted parentheses are given below:

    a. If $A$ is an atomic formula, then the outermost parentheses in $A$ are omitted. These parentheses are added if $A$ is used to obtain $(\neg A)$, $(\forall a)A$, $(\exists a)A$ and $(\exists! a)A$.
    b. The outermost parentheses in $(\neg A)$, $(A \vee B)$, $(A \wedge B)$, $(A \Rightarrow B)$ and $(A \Leftrightarrow B)$ are omitted. The outermost parentheses are added to the formula $\neg A$ to obtain the formulas $(\neg A) \vee C$, $(\neg A) \wedge C$, $(\neg A) \Rightarrow C$ and $(\neg A) \Leftrightarrow C$, where $C$ is a formula. The outermost parentheses are added to $A \vee B$, $A \wedge B$, $A \Rightarrow B$ and $A \Leftrightarrow B$ to obtain the formulas $\neg(A \vee B)$, $\neg(A \wedge B)$, $\neg(A \Rightarrow B)$, $\neg(A \Leftrightarrow B)$, $(\forall a)(A \vee B)$, $(\forall a)(A \wedge B)$, $(\forall a)(A \Rightarrow B)$, $(\forall a)(A \Leftrightarrow B)$, $(\exists a)(A \vee B)$, $(\exists a)(A \wedge B)$, $(\exists a)(A \Rightarrow B)$, $(\exists a)(A \Leftrightarrow B)$, $(\exists! a)(A \vee B)$, $(\exists! a)(A \wedge B)$, $(\exists! a)(A \Rightarrow B)$, $(\exists! a)(A \Leftrightarrow B)$, $(A \vee B) \vee C$, $C \vee (A \vee B)$, $(A \vee B) \wedge C$, $C \wedge (A \vee B)$, $(A \vee B) \Rightarrow C$, $C \Rightarrow (A \vee B)$, $(A \vee B) \Leftrightarrow C$, $C \Leftrightarrow (A \vee B)$, $(A \wedge B) \vee C$, $C \vee (A \wedge B)$, $(A \wedge B) \wedge C$, $C \wedge (A \wedge B)$,

$(A \wedge B) \Rightarrow C, C \Rightarrow (A \wedge B), (A \wedge B) \Leftrightarrow C, C \Leftrightarrow (A \wedge B), (A \Rightarrow B) \vee C,$ $C \vee (A \Rightarrow B), (A \Rightarrow B) \wedge C, C \wedge (A \Rightarrow B), (A \Rightarrow B) \Rightarrow C, C \Rightarrow (A \Rightarrow B),$ $(A \Rightarrow B) \Leftrightarrow C, C \Leftrightarrow (A \Rightarrow B), (A \Leftrightarrow B) \vee C, C \vee (A \Leftrightarrow B), (A \Leftrightarrow B) \wedge C,$ $C \wedge (A \Leftrightarrow B), (A \Leftrightarrow B) \Rightarrow C, C \Rightarrow (A \Leftrightarrow B), (A \Leftrightarrow B) \Leftrightarrow C$ and $C \Leftrightarrow (A \Leftrightarrow B)$ where $C$ is a formula.

4. Quantifier: $(\forall a)$, $(\exists a)$ and $(\exists! a)$ are called quantifiers. $(\forall a)$ is called a universal quantifier. $(\exists a)$ is called an existential quantifier. $(\exists! a)$ is called a uniqueness quantifier.
5. Scope of a quantifier: In $(\forall a)A$, $(\exists a)A$ and $(\exists! a)A$, $A$ is called the scope of the quantifiers $(\forall a)$, $(\exists a)$ and $(\exists! a)$, respectively.
6. Bound occurrence of a variable: An occurrence of a variable $a$ is said to be bound in a formula $A$ if either it is the occurrence of $a$ in a quantifier ($(\forall a)$, $(\exists a)$ or $(\exists! a)$) in $A$, or it is the occurrence of $a$ in the scope of a quantifier, $(\forall a)$, $(\exists a)$ or $(\exists! a)$, in $A$.
7. Free occurrence of a variable: An occurrence of a variable $a$ in a formula $A$ is said to be free if it is not a bound occurrence.
8. Bound variable: A variable $a$ is said to be bound in a formula $A$ if $A$ has a bound occurrence of $a$.
9. Free variable: A variable $a$ is said to be free in a formula $A$ if $A$ has a free occurrence of $a$. $A(a_1, a_2, ..., a_n)$ denotes a formula $A$ in which the free variables are $a_1, a_2, ..., a_n$, where $n$ is a positive integer. $A(a, a_1, a_2, ..., a_n)$ denotes a formula $A$ in which the free variables are $a, a_1, a_2, ..., a_n$, where $n$ is a non-negative integer.
10. Closed formula: A closed formula is a formula with no free variables.
11. Axiom: There are two types of axioms: logical axioms and proper axioms. The logical axioms are given in the form of rules in Sect. 1.4. The proper axioms are the axioms specific to a theory such as Zermelo set theory. The proper axioms are closed formulas. Section 1.3 gives the details of proper axioms.
12. Theorem: A theorem is a closed formula that needs to be proved.
13. Proof of a theorem: A proof of a theorem is a sequence of formulas called steps that are numbered 1, 2, 3, . . . . There is, however, an exception where the step is not a formula; this exception is given in Sect. 1.4. The last step of the proof is the theorem itself. For each step, the rule, based on which the step is obtained, is indicated in the proof; abbreviations are used to indicate the rules. The rules that apply to obtain the steps in a proof are given in Sect. 1.4.
14. Subproof: A subproof is a certain set of consecutive steps in the proof of a theorem. Section 1.4 gives the details of which set of consecutive steps in a proof constitute a subproof.

## 1.3 Proper Axiom

The proper axioms can be grouped into four types:

1. Main proper axioms: The main proper axioms are foundational statements involving undefined predicates, undefined operations and undefined constants. In Zermelo set theory [1, 4, 6], the main proper axioms are called Zermelo axioms. The Zermelo axioms contain an undefined predicate $\in$ with two arguments; the Zermelo axioms do not contain an undefined operation or an undefined constant. If $\alpha$ and $\beta$ are terms, then $\alpha \in \beta$ is an atomic formula. The set of Zermelo axioms consists of six axioms and an axiom schema. The following examples show how a main proper axiom (for example, a Zermelo axiom and the Zermelo axiom schema) is stated.

   **Axiom 2.1 (Zermelo Axiom: Axiom of Extensionality)**
   $(\forall a)(\forall b)((\forall c)(c \in a \Leftrightarrow c \in b) \Rightarrow a = b)$

   **Axiom Schema 2.1 (Zermelo Axiom Schema: Axiom Schema of Separation)**

   $$(\forall a)(\forall a_1)(\forall a_2)...(\forall a_n)(\exists b)(\forall c)(c \in b \Leftrightarrow (c \in a \wedge A(c, a_1, a_2, ..., a_n)))$$

   Axiom 2.1 is a main proper axiom, and Axiom Schema 2.1 is an axiom schema. In Axiom Schema 2.1, $n$ is a non-negative integer; if $n$ is zero, the axiom schema is $(\forall a)(\exists b)(\forall c)(c \in b \Leftrightarrow (c \in a \wedge A(c)))$. An axiom schema stands for an infinite number of axioms, which are called instances of the axiom schema. In an instance of the axiom schema, a specific value is chosen for $n$ and a specific formula is chosen for $A(c, a_1, a_2, ..., a_n)$. For example, $(\forall a)(\forall a_1)(\exists b)(\forall c)(c \in b \Leftrightarrow (c \in a \wedge c \in a_1))$ is an instance of Axiom Schema 2.1.
2. Definitions of predicates [7]: These axioms are definitions of predicates other than "=" and the undefined predicates in the main proper axioms. The following example shows how a predicate is defined.

   **Axiom 2.5 (Definition of Predicate Which If Applied to the Arguments $\alpha$ and $\beta$ Gives the Atomic Formula "$\alpha \neq \beta$")**
   $(\forall a)(\forall b)(a \neq b \Leftrightarrow \neg(a = b))$
   The predicate should not have been already defined as a predicate or an operation with the same number of arguments.
   Two or more predicates can mean the same thing as in the following example.

   **Axiom 2.9 (Definitions of Predicates Which If Applied to the Arguments $\alpha$ and $\beta$ Give the Atomic Formulas "$\alpha \subseteq \beta$" and "$\alpha$ is a subset of $\beta$")**
   $(\forall a)(\forall b)(a \subseteq b \Leftrightarrow (\forall c)(c \in a \Rightarrow c \in b))$
   $(\forall a)(\forall b)(a \text{ is a subset of } b \Leftrightarrow (\forall c)(c \in a \Rightarrow c \in b))$
3. Definitions of operations [1]: These axioms are definitions of operations other than the undefined operations in the main proper axioms. The definition of an operation should be preceded by a theorem which has a uniqueness quantifier. The following example shows a theorem followed by an axiom that defines an operation.

   **Theorem 2.2** $(\forall a)(\forall b)(\exists! c)(\forall d)(d \in c \Leftrightarrow (d = a \vee d = b))$

**Axiom 2.3 (Definition of Operation, from Theorem 2.2, Which If Applied to the Arguments $\alpha$ and $\beta$ Gives the Term "$\{\alpha, \beta\}$")**

$(\forall a)(\forall b)(\forall c)(c \in \{a, b\} \Leftrightarrow (c = a \vee c = b))$

In the above example, Axiom 2.3 is obtained from Theorem 2.2 in two steps:

a. "$(\exists! c)$" is removed and all occurrences of $c$ are replaced with $\{a, b\}$ to obtain $(\forall a)(\forall b)(\forall d)(d \in \{a, b\} \Leftrightarrow (d = a \vee d = b))$.
b. All occurrences of $d$ are replaced with $c$. Since $c$ does not occur in $(\forall a)(\forall b)(\forall d)(d \in \{a, b\} \Leftrightarrow (d = a \vee d = b))$, all occurrences of $d$ can be replaced with $c$.

The operation should not have been already defined as a predicate or an operation with the same number of arguments. However, for example, if an operation, which if applied to the argument $\alpha$ gives $\cup\,\alpha$, is already defined, then a new operation, which if applied to the arguments $\alpha$ and $\beta$ gives $\alpha \cup \beta$, can be defined.

The number of arguments of the operation is equal to one less than the number of free variables in the scope of the uniqueness quantifier, in the theorem, that is being removed to obtain the axiom. In Theorem 2.2, there are three free variables ($a$, $b$ and $c$) in the scope of $(\exists! c)$; therefore, the number of arguments of the operation is two.

Two or more operations can mean the same thing as in the following example.

**Theorem 2.16** $(\forall a)(\exists! b)(\forall c)(c \in b \Leftrightarrow (\exists d)(c \in d \wedge d \in a))$

**Axiom 2.7 (Definitions of Operations, from Theorem 2.16, Which If Applied to the Argument $\alpha$ Give the Terms "$\cup\,\alpha$", "sum set of $\alpha$" and "union of elements of $\alpha$")**

$(\forall a)(\forall b)(b \in \cup a \Leftrightarrow (\exists c)(b \in c \wedge c \in a))$
$(\forall a)(\forall b)(b \in \text{sum set of } a \Leftrightarrow (\exists c)(b \in c \wedge c \in a))$
$(\forall a)(\forall b)(b \in \text{union of elements of } a \Leftrightarrow (\exists c)(b \in c \wedge c \in a))$

4. Definitions of constants [1]: These axioms are definitions of constants other than the undefined constants in the main proper axioms. The definition of a constant should be preceded by a theorem which has a uniqueness quantifier. The following example shows a theorem followed by an axiom that defines a constant.

**Theorem 2.123** $(\exists! a)(\forall b)(b \in a \Leftrightarrow (b \in \mathrm{N} \times \mathrm{N} \wedge (\exists c)(b = (c, c \cup \{c\}))))$

**Axiom 2.32 (Definition of Constant "successor" from Theorem 2.123)**

$(\forall a)(a \in \text{successor} \Leftrightarrow (a \in \mathrm{N} \times \mathrm{N} \wedge (\exists b)(a = (b, b \cup \{b\}))))$

In the above example, Axiom 2.32 is obtained from Theorem 2.123 in three steps:

a. "$(\exists! a)$" is removed and all occurrences of $a$ are replaced with "successor" to obtain $(\forall b)(b \in \text{successor} \Leftrightarrow (b \in \mathrm{N} \times \mathrm{N} \wedge (\exists c)(b = (c, c \cup \{c\}))))$.
b. All occurrences of $b$ are replaced with $a$; since $a$ does not occur in $(\forall b)(b \in \text{successor} \Leftrightarrow (b \in \mathrm{N} \times \mathrm{N} \wedge (\exists c)(b = (c, c \cup \{c\}))))$, all occurrences of $b$ can

be replaced with $a$ to get $(\forall a)(a \in \text{successor} \Leftrightarrow (a \in \text{N} \times \text{N} \wedge (\exists c)(a = (c, c \cup \{c\}))))$.

c. All occurrences of $c$ are replaced with $b$; since $b$ does not occur in $(\forall a)(a \in \text{successor} \Leftrightarrow (a \in \text{N} \times \text{N} \wedge (\exists c)(a = (c, c \cup \{c\}))))$, all occurrences of $c$ can be replaced with $b$.

The constant should not have been already defined as a constant. The number of free variables in the scope of the uniqueness quantifier, in the theorem, that is being removed to obtain the axiom, is one.

Two or more constants can mean the same thing as in the following example.

**Theorem 2.41** $(\exists!a)(\forall b)(b \notin a)$

**Axiom 2.14 (Definitions of Constants "$\phi$" and "empty set" from Theorem 2.41)**

$(\forall a)(a \notin \phi)$
$(\forall a)(a \notin \text{empty set})$

## 1.4 Proof

For each step in a proof, the rule, based on which the step is obtained, is indicated in the proof; abbreviations are used to indicate the rules. The steps in a proof are obtained from the following rules [2, 4, 5, 8].

1. A previously stated proper axiom, an instance of a previously stated proper axiom schema or a previously proved theorem, can be a step in the proof. All proper axioms, instances of proper axiom schemas and theorems used in proving a theorem, should appear as steps at the beginning of the proof. In a proof, the rule for including a proper axiom is abbreviated as Ax followed by its number; for example, "Ax 2.1" stands for "Axiom 2.1". In a proof, the rule for including an instance of a proper axiom schema is abbreviated as AS followed by its number; for example, "AS 2.1" stands for "an instance of Axiom Schema 2.1". In a proof, the rule for including a theorem is abbreviated as T followed by its number; for example, "T 2.1" stands for "Theorem 2.1". While including a previously stated proper axiom, an instance of a previously stated proper axiom schema or a previously proved theorem as a step in the proof, all occurrences of a variable in the axiom, instance of axiom schema or theorem, can be replaced with another variable which is not there in this axiom, instance of axiom schema or theorem. For example, if $(\forall a)(a \notin \phi)$ is a previously stated axiom, then $(\forall b)(b \notin \phi)$ can be a step in the proof. There can be multiple such replacements of variables. For example, if $(\forall a)((\forall b)(b \notin a) \Rightarrow (\forall c)(a \subseteq c))$ is a previously stated theorem, then $(\forall c)((\forall b)(b \notin c) \Rightarrow (\forall a)(c \subseteq a))$ can be a step in the proof, which is obtained by first replacing $c$ with say, $d$, then replacing $a$ with $c$ and finally replacing $d$ with $a$.

2. The rules for omission of outermost parentheses and addition of the omitted parentheses to the formulas if these formulas are used to construct certain other formulas, given in Sect. 1.2, are applicable to the steps in the proof.
3. If $\alpha$ and $\beta$ are terms, $\alpha = \beta$ is a previous step in the proof and a previous step in the proof is a formula $A$ containing $\alpha$, then this formula $A$ with some or all occurrences of $\alpha$ replaced with $\beta$ can be a step in the proof. Similarly, if $\alpha$ and $\beta$ are terms, $\alpha = \beta$ is a previous step in the proof and a previous step in the proof is a formula $A$ containing $\beta$, then this formula $A$ with some or all occurrences of $\beta$ replaced with $\alpha$ can be a step in the proof. This rule is called "Identity Substitution", abbreviated as IS. In the proof, the rule for this step is indicated as IS followed by two numbers; these two numbers are the step numbers of the two previous steps, one being $\alpha = \beta$ and the other is the formula $A$ containing $\alpha$ or $\beta$.
4. A formula obtained by a "Quantifier Negation" rule can be a step in the proof. These rules are of the form $A \equiv B$. If $A$ is a previous step in the proof, then using this rule, $B$ can be a step in the proof and the rule is indicated by the abbreviation QN followed by the step number of $A$; also, if $B$ is a previous step in the proof, then using this rule, $A$ can be a step in the proof and the rule is indicated by QN followed by the step number of $B$. The following are the "Quantifier Negation" rules.

   a. $\neg(\forall a)A \equiv (\exists a)\neg A$
   b. $\neg(\exists a)A \equiv (\forall a)\neg A$
   c. $\neg(\forall a)\neg A \equiv (\exists a)A$
   d. $\neg(\exists a)\neg A \equiv (\forall a)A$

5. A formula containing the uniqueness quantifier can be written as a formula containing existential and universal quantifiers; this is stated as a rule of the form $A \equiv B$. A formula obtained by this rule can be a step in the proof. By this rule, if $A$ is a previous step in the proof, then $B$ can be a step in the proof; the rule is indicated by the abbreviation UQ followed by the step number of $A$. Also, if $B$ is a previous step in the proof, then $A$ can be a step in the proof, and the rule is indicated by UQ followed by the step number of $B$. This rule is given by the following four forms:

   a. $(\exists !a)A(a, a_1, a_2, ..., a_n) \equiv$
   $(\exists a)(A(a, a_1, a_2, ..., a_n) \wedge (\forall b)(A(b, a_1, a_2, ..., a_n) \Rightarrow b = a))$
   b. $(\exists !a)A(a, a_1, a_2, ..., a_n) \equiv$
   $(\exists a)(A(a, a_1, a_2, ..., a_n) \wedge (\forall b)(A(b, a_1, a_2, ..., a_n) \Rightarrow a = b))$
   c. $(\exists !a)A(a, a_1, a_2, ..., a_n) \equiv (\exists a)A(a, a_1, a_2, ..., a_n)$
   $\wedge (\forall b)(\forall c)((A(b, a_1, a_2, ..., a_n) \wedge A(c, a_1, a_2, ..., a_n)) \Rightarrow b = c)$
   d. $(\exists !a)A(a, a_1, a_2, ..., a_n) \equiv (\exists a)A(a, a_1, a_2, ..., a_n)$
   $\wedge (\forall b)(\forall c)((A(b, a_1, a_2, ..., a_n) \wedge A(c, a_1, a_2, ..., a_n)) \Rightarrow c = b)$

6. Let $(\exists a)A(a, a_1, a_2, ..., a_n)$ be a step in the proof. A formula $A(b, a_1, a_2, ..., a_n)$, obtained from $(\exists a)A(a, a_1, a_2, ..., a_n)$ by removing $(\exists a)$ and replacing all occurrences of $a$ in $A(a, a_1, a_2, ..., a_n)$ with $b$, can be a step in the proof.

Obtaining a step in this way is called Existential Instantiation, abbreviated as EI. The variable $b$ is "flagged" in the same step. The flagged variable should be new to the proof; in other words, the flagged variable should not have appeared before this step. Also, the flagged variable should not be a variable in the theorem which is being proved. In the proof, the rule for this step is indicated as "EI" followed by the step number, of $(\exists a)A(a, a_1, a_2, ..., a_n)$, which is followed by "flag $b$"; for example, if the step number of $(\exists a)A(a, a_1, a_2, ..., a_n)$ is 10, then the rule for obtaining the step $A(b, a_1, a_2, ..., a_n)$ is indicated as "EI 10, flag $b$", and the variable $b$ is said to be flagged.

7. Let $A(b, a_1, a_2, ..., a_n)$ be a step in the proof. A formula $(\forall a)A(a, a_1, a_2, ..., a_n)$, obtained from $A(b, a_1, a_2, ..., a_n)$ by adding $(\forall a)$ and replacing all occurrences of $b$ in $A(b, a_1, a_2, ..., a_n)$ with $a$, can be a step in a proof. Obtaining a step in this way is called Universal Generalization, abbreviated as UG. The step of Universal Generalization follows immediately after a subproof; in the proof, the rule for this step is indicated as UG followed by the range of step numbers of the subproof. The first step of this subproof is "flag $b$"; this step is called Flagging Step for Universal Generalization and the rule is abbreviated as FSUG, and $b$ is said to be a flagged variable. The flagged variable must be new to the proof; in other words, the flagged variable should not have appeared before the first step of the subproof. Also, the flagged variable should not be a variable in the theorem which is being proved. The last step in the subproof is $A(b, a_1, a_2, ..., a_n)$. To obtain $(\forall a)A(a, a_1, a_2, ..., a_n)$ from $A(b, a_1, a_2, ..., a_n)$, $a$ should not be a flagged variable and $a$ should not be a variable in $A(b, a_1, a_2, ..., a_n)$.
8. Let $(\forall a)A(a, a_1, a_2, ..., a_n)$ be a step in the proof. A formula obtained from $(\forall a)A(a, a_1, a_2, ..., a_n)$ by removing $(\forall a)$ and replacing all occurrences of $a$ in $A(a, a_1, a_2, ..., a_n)$ with a term $\alpha$ can be a step in the proof if all variables in $\alpha$ are flagged variables that are flagged in previous steps; a variable is flagged in a step, the rule for which is abbreviated as FSUG or EI. Obtaining a step in this way is called Universal Instantiation, abbreviated as UI. In the proof, the rule for this step is indicated as UI followed by the step number of the step $(\forall a)A(a, a_1, a_2, ..., a_n)$.
9. Let $A$ be a step in the proof. If $a$ is not a flagged variable and is not a variable in $A$ and $\alpha$ is a term in $A$ such that all variables in $\alpha$ are flagged variables, then the formula $(\exists a)A$, obtained from $A$ by adding $(\exists a)$ and replacing some or all occurrences of $\alpha$ with $a$, can be a step in the proof. Obtaining a step in this way is called Existential Generalization, abbreviated as EG. The rule for this step is indicated as EG followed by the step number of $A$.
10. A formula obtained by a rule of inference can be a step in the proof. A rule of inference consists of a set of formulas called the premises followed by a formula called the conclusion; the conclusion is preceded by the symbol "$\therefore$". For example, a rule of inference can consist of premises $A$, $B$ and $C$ followed by $\therefore D$, where $D$ is the conclusion. If all premises of a rule of inference are previous steps in the proof, then the conclusion of the rule of inference can be a step in the proof. The rules of inference, along with their

abbreviations, are given in Table 1.2. In the proof, a rule of inference is indicated by its abbreviation followed by the step numbers of the premises of the rule of inference. For example, "Modus Ponens" can be used to obtain a step of the form $B$, if there are two steps in the proof which are of the form $A \Rightarrow B$ and $A$; in the proof, the rule is indicated by its abbreviation MP followed by the step numbers of $A \Rightarrow B$ and $A$. If "Addition" is used to obtain a step in the proof, then all free variables in "$B$", in the conclusion, should be flagged variables, and a flagged variable cannot be a bound variable in "$B$".

11. A formula obtained by a replacement rule can be a step in the proof. A replacement rule is of the form $A \equiv B$, where $A$ and $B$ are formulas. If $A$ is a previous step in the proof, then using the replacement rule $A \equiv B$, $B$ can be a step in the proof; also, if $B$ is a previous step in the proof, then using the replacement rule $A \equiv B$, $A$ can be a step in the proof. The replacement rules and their abbreviations are given in Table 1.3. A replacement rule is used to obtain a new step from a previous step. In the proof, the replacement rule is indicated by its abbreviation followed by the step number of the step from which the new step is obtained. For example, if "Double Negation" is used to obtain a step, then in the proof, the rule is indicated by its abbreviation DN followed by the step number of the step from which the new step is obtained. A replacement rule can be used on a formula which is a part of a previous step. For example, "Double Negation" can be used to obtain the step $A \Rightarrow \neg\neg B$ if $A \Rightarrow B$ is a previous step in the proof.
12. If $A$ and $B$ are formulas, then $A \Rightarrow B$ is called a conditional; $A$ is called antecedent and $B$ is called consequent. A Conditional Proof is a subproof in which the first step is antecedent $A$ and the last step is consequent $B$; the rule for the first step is indicated as ACP, which is the abbreviation of Assumption for Conditional Proof. All free variables in the antecedent should be flagged variables, and a flagged variable should not be a bound variable in the antecedent. The step immediately after the consequent is $A \Rightarrow B$; the rule for this step is indicated as CP (which is the abbreviation of Conditional Proof) followed by the range of step numbers of the subproof.
13. An Indirect Proof is a subproof. If $A$ is a formula, an Indirect Proof proves $\neg A$. The first step of the subproof is $A$, the rule for which is indicated as AIP (abbreviation of Assumption for Indirect Proof); all free variables in $A$ should be flagged variables, and a flagged variable should not be a bound variable in $A$. The last step of the subproof is of the form $B \wedge \neg B$, where $B$ is a formula. The step immediately after the subproof is $\neg A$; the rule for this step is indicated as IP (which is the abbreviation of Indirect Proof) followed by the range of step numbers of the subproof.
14. A flagged variable should not appear outside the subproof in which it gets flagged.
15. All steps in a subproof which arises in the cases of Universal Generalization (rule 7), Conditional Proof (rule 12) and Indirect Proof (rule 13) are indented. A step in a subproof cannot be used to obtain a step outside the subproof except

**Table 1.2** Rules of inference

| Name and abbreviation | Rule | | | |
|---|---|---|---|---|
| Modus Ponens, MP | $A \Rightarrow B$<br>$A$<br>$\therefore B$ | | | |
| Modus Tollens, MT | $A \Rightarrow B$<br>$\neg B$<br>$\therefore \neg A$ | | | |
| Hypothetical Syllogism, HS | $A \Rightarrow B$<br>$B \Rightarrow C$<br>$\therefore A \Rightarrow C$ | | | |
| Simplification, S | $A \wedge B$<br>$\therefore A$ | $A \wedge B$<br>$\therefore B$ | | |
| Dilemma, D | $A \Rightarrow B$<br>$C \Rightarrow D$<br>$A \vee C$<br>$\therefore B \vee D$ | $A \Rightarrow B$<br>$C \Rightarrow D$<br>$A \vee C$<br>$\therefore D \vee B$ | $A \Rightarrow B$<br>$A \vee C$<br>$\therefore B \vee C$ | $A \Rightarrow B$<br>$A \vee C$<br>$\therefore C \vee B$ |
| | $A \Rightarrow B$<br>$C \vee A$<br>$\therefore B \vee C$ | $A \Rightarrow B$<br>$C \vee A$<br>$\therefore C \vee B$ | $A \Leftrightarrow B$<br>$C \Rightarrow D$<br>$A \vee C$<br>$\therefore B \vee D$ | $A \Leftrightarrow B$<br>$C \Rightarrow D$<br>$A \vee C$<br>$\therefore D \vee B$ |
| | $A \Leftrightarrow B$<br>$C \Rightarrow D$<br>$C \vee A$<br>$\therefore B \vee D$ | $A \Leftrightarrow B$<br>$C \Rightarrow D$<br>$C \vee A$<br>$\therefore D \vee B$ | $A \Leftrightarrow B$<br>$C \Rightarrow D$<br>$B \vee C$<br>$\therefore A \vee D$ | $A \Leftrightarrow B$<br>$C \Rightarrow D$<br>$B \vee C$<br>$\therefore D \vee A$ |
| | $A \Leftrightarrow B$<br>$C \Rightarrow D$<br>$C \vee B$<br>$\therefore A \vee D$ | $A \Leftrightarrow B$<br>$C \Rightarrow D$<br>$C \vee B$<br>$\therefore D \vee A$ | $A \Leftrightarrow B$<br>$A \vee C$<br>$\therefore B \vee C$ | $A \Leftrightarrow B$<br>$A \vee C$<br>$\therefore C \vee B$ |
| | $A \Leftrightarrow B$<br>$C \vee A$<br>$\therefore B \vee C$ | $A \Leftrightarrow B$<br>$C \vee A$<br>$\therefore C \vee B$ | $A \Leftrightarrow B$<br>$B \vee C$<br>$\therefore A \vee C$ | $A \Leftrightarrow B$<br>$B \vee C$<br>$\therefore C \vee A$ |
| | $A \Leftrightarrow B$<br>$C \vee B$<br>$\therefore A \vee C$ | $A \Leftrightarrow B$<br>$C \vee B$<br>$\therefore C \vee A$ | $A \Leftrightarrow B$<br>$C \Leftrightarrow D$<br>$A \vee C$<br>$\therefore B \vee D$ | $A \Leftrightarrow B$<br>$C \Leftrightarrow D$<br>$A \vee C$<br>$\therefore D \vee B$ |
| | $A \Leftrightarrow B$<br>$C \Leftrightarrow D$<br>$A \vee D$<br>$\therefore B \vee C$ | $A \Leftrightarrow B$<br>$C \Leftrightarrow D$<br>$A \vee D$<br>$\therefore C \vee B$ | $A \Leftrightarrow B$<br>$C \Leftrightarrow D$<br>$D \vee A$<br>$\therefore B \vee C$ | $A \Leftrightarrow B$<br>$C \Leftrightarrow D$<br>$D \vee A$<br>$\therefore C \vee B$ |
| | $A \Leftrightarrow B$<br>$C \Leftrightarrow D$<br>$B \vee D$<br>$\therefore A \vee C$ | $A \Leftrightarrow B$<br>$C \Leftrightarrow D$<br>$B \vee D$<br>$\therefore C \vee A$ | | |

(continued)

**Table 1.2** (continued)

| Name and abbreviation | Rule | | | |
|---|---|---|---|---|
| Conjunction, C | $A$<br>$B$<br>$\therefore A \wedge B$ | | | |
| Disjunctive Syllogism, DS | $A \vee B$<br>$\neg A$<br>$\therefore B$ | $A \vee B$<br>$\neg B$<br>$\therefore A$ | | |
| Addition, Ad | $A$<br>$\therefore A \vee B$ | $A$<br>$\therefore B \vee A$ | | |
| Biconditional Elimination, BEl | $A \Leftrightarrow B$<br>$A$<br>$\therefore B$ | $A \Leftrightarrow B$<br>$B$<br>$\therefore A$ | | |
| Negative Biconditional Elimination, NBE | $A \Leftrightarrow B$<br>$\neg A$<br>$\therefore \neg B$ | $A \Leftrightarrow B$<br>$\neg B$<br>$\therefore \neg A$ | | |
| Biconditional Transitivity, BT | $A \Leftrightarrow B$<br>$B \Leftrightarrow C$<br>$\therefore A \Leftrightarrow C$ | $A \Leftrightarrow B$<br>$B \Leftrightarrow C$<br>$\therefore C \Leftrightarrow A$ | $A \Leftrightarrow B$<br>$C \Leftrightarrow B$<br>$\therefore A \Leftrightarrow C$ | $A \Leftrightarrow B$<br>$A \Leftrightarrow C$<br>$\therefore B \Leftrightarrow C$ |
| Identity Transitivity, IT | $\alpha = \beta$<br>$\beta = \gamma$<br>$\therefore \alpha = \gamma$ | $\alpha = \beta$<br>$\beta = \gamma$<br>$\therefore \gamma = \alpha$ | $\alpha = \beta$<br>$\gamma = \beta$<br>$\therefore \alpha = \gamma$ | $\alpha = \beta$<br>$\alpha = \gamma$<br>$\therefore \beta = \gamma$ |

that the entire subproof is used to obtain the step immediately after the subproof as mentioned in rules 7, 12 and 13. There can be a subproof within a subproof.

16. If $A$ is a formula, then the formula $A \Rightarrow A$ is called Principle of Identity. If $A$ is a formula such that all free variables in $A$ are flagged variables and a flagged variable is not a bound variable in $A$, then the formula $A \Rightarrow A$ can be a step in the proof. The rule for this step is abbreviated as I.
17. If $A$ is a formula, then the formula $A \vee \neg A$ is called Principle of Excluded Middle. If $A$ is a formula such that all free variables in $A$ are flagged variables and a flagged variable is not a bound variable in $A$, then the formula $A \vee \neg A$ can be a step in the proof. The rule for this step is abbreviated as EM.
18. If $\alpha$ is a term in which all variables are flagged variables, then $\alpha = \alpha$ can be a step in the proof. This rule is called Identity Reflexivity, abbreviated as IR.
19. If $(\forall a)(\forall b)A$ is a previous step in the proof, then $(\forall b)(\forall a)A$ can be a step in the proof. The rule for obtaining this step is indicated as IQV, which is the abbreviation of "Interchange of Quantifier Variables", followed by the step number of $(\forall a)(\forall b)A$. If $(\exists a)(\exists b)A$ is a previous step in the proof, then $(\exists b)(\exists a)A$ can be a step in the proof. The rule for obtaining this step is indicated as IQV followed by the step number of $(\exists a)(\exists b)A$.

It can be seen that all steps in the proof, except the step to which the rule abbreviated as FSUG applies, are formulas. It can also be seen that in any step in a proof, other

**Table 1.3** Replacement rules

| Name and abbreviation | Rule |
|---|---|
| Double Negation, DN | $A \equiv \neg\neg A$ |
| Duplication, Du | $A \equiv A \vee A$ |
| | $A \equiv A \wedge A$ |
| Commutation, Co | $A \vee B \equiv B \vee A$ |
| | $A \wedge B \equiv B \wedge A$ |
| | $A \Leftrightarrow B \equiv B \Leftrightarrow A$ |
| Association, A | $(A \vee B) \vee C \equiv A \vee (B \vee C)$ |
| | $(A \wedge B) \wedge C \equiv A \wedge (B \wedge C)$ |
| Contraposition, Con | $A \Rightarrow B \equiv (\neg B) \Rightarrow \neg A$ |
| De Morgan's rules, DM | $\neg(A \vee B) \equiv (\neg A) \wedge \neg B$ |
| | $\neg(A \wedge B) \equiv (\neg A) \vee \neg B$ |
| Biconditional Exchange, BE | $A \Leftrightarrow B \equiv (A \Rightarrow B) \wedge (B \Rightarrow A)$ |
| | $B \Leftrightarrow A \equiv (A \Rightarrow B) \wedge (B \Rightarrow A)$ |
| Distribution, Di | $A \wedge (B \vee C) \equiv (A \wedge B) \vee (A \wedge C)$ |
| | $A \vee (B \wedge C) \equiv (A \vee B) \wedge (A \vee C)$ |
| Conditional Exchange, CE | $A \Rightarrow B \equiv (\neg A) \vee B$ |
| Exportation, E | $(A \wedge B) \Rightarrow C \equiv A \Rightarrow (B \Rightarrow C)$ |
| Identity Symmetry, ISy | $\alpha = \beta \equiv \beta = \alpha$ |

than the one to which the rule abbreviated as FSUG applies, all free variables are flagged variables.

The following is an example of a theorem and its proof.

**Theorem 2.1** $(\forall a)(\forall b)(a = b \Rightarrow (\forall c)(c \in a \Leftrightarrow c \in b))$

***Proof***

| | | |
|---|---|---|
| 1. | flag $d$ | FSUG |
| 2. | flag $e$ | FSUG |
| 3. | flag $f$ | FSUG |
| 4. | $f \in d \Rightarrow f \in d$ | I |
| 5. | $(f \in d \Rightarrow f \in d) \wedge (f \in d \Rightarrow f \in d)$ | Du 4 |
| 6. | $f \in d \Leftrightarrow f \in d$ | BE 5 |
| 7. | $(\forall c)(c \in d \Leftrightarrow c \in d)$ | UG 3-6 |
| 8. | $d = e$ | ACP |
| 9. | $(\forall c)(c \in d \Leftrightarrow c \in e)$ | IS 7, 8 |
| 10. | $d = e \Rightarrow (\forall c)(c \in d \Leftrightarrow c \in e)$ | CP 8-9 |
| 11. | $(\forall b)(d = b \Rightarrow (\forall c)(c \in d \Leftrightarrow c \in b))$ | UG 2-10 |
| 12. | $(\forall a)(\forall b)(a = b \Rightarrow (\forall c)(c \in a \Leftrightarrow c \in b))$ | UG 1-11 |

□

The last step of the proof is the theorem which is proved. There are four subproofs in this proof namely:

1. Step 1 to step 11
2. Step 2 to step 10
3. Step 3 to step 6
4. Step 8 to step 9

The first three subproofs are subproofs for Universal Generalization. The last subproof (from step 8 to step 9) is a Conditional Proof. The subproof from step 2 to step 10 is within the subproof from step 1 to step 11. The subproof from step 3 to step 6 and that from step 8 to step 9 are subproofs within the subproof from step 2 to step 10. The steps in the subproof from step 1 to step 11 are indented; the steps in the subproof from step 2 to step 10 are indented further since these are steps in the subproof within the subproof from step 1 to step 11. The first step (step 1) of the subproof from step 1 to step 11 is a flagging step in which the variable $d$ is flagged; the last step (step 11) of this subproof contains $d$ as a free variable, and the Universal Generalization of this step is the next step (step 12). $a$, $b$ or $c$ cannot be the flagged variable in step 1 since these variables are in the theorem that is being proved. The subproof from step 8 to step 9 is a Conditional Proof; the first and last steps (steps 8 and 9, respectively) of this subproof are the antecedent and the consequent, respectively, of the conditional that is proved; the conditional that is proved is the step (step 10) immediately after the subproof that gives the Conditional Proof.

## References


1. Mendelson, E.: Introduction to Mathematical Logic, 3rd edn. Wadsworth, Pacific Grove (1987)
2. Copi, I.M., Cohen, C.: Introduction to Logic, 9th edn. Prentice-Hall of India, New Delhi (1995)
3. Margaris, A.: First Order Mathematical Logic. Dover, New York (1990)
4. Mendelson, E.: Number Systems and the Foundations of Analysis. Dover, New York (2008)
5. Klenk, V.: Understanding Symbolic Logic, 5th edn. Pearson Education, Upper Saddle River (2009)
6. Jech, T.: Set Theory. Springer, New York (2002)
7. Shoenfield, J.R.: Mathematical Logic. Addison-Wesley, Reading (1967)
8. Jongsma, C.: Introduction to Discrete Mathematics via Logic and Proof. Springer, New York (2019)

# Chapter 2
# Set Theory

This chapter gives formal proofs of theorems in elementary set theory [1]. Zermelo axiom system [1–3] is chosen as the basis for the set theory. Proving the existence of a Peano system is one of the objectives.

The Zermelo axioms contain an undefined predicate $\in$ with two arguments. If $\alpha$ and $\beta$ are terms, then $\alpha \in \beta$ is an atomic formula. In Zermelo set theory, the terms are thought of intuitively as sets. "$\alpha \in \beta$" can be thought of intuitively as "$\alpha$ is a member of $\beta$", "$\alpha$ is an element of $\beta$" or "$\alpha$ belongs to $\beta$".

## 2.1 Equality of Sets and Pairing

**Theorem 2.1** $(\forall a)(\forall b)(a = b \Rightarrow (\forall c)(c \in a \Leftrightarrow c \in b))$

***Proof***

| | | |
|---|---|---|
| 1. | flag $d$ | FSUG |
| 2. | flag $e$ | FSUG |
| 3. | flag $f$ | FSUG |
| 4. | $f \in d \Rightarrow f \in d$ | I |
| 5. | $(f \in d \Rightarrow f \in d) \wedge (f \in d \Rightarrow f \in d)$ | Du 4 |
| 6. | $f \in d \Leftrightarrow f \in d$ | BE 5 |
| 7. | $(\forall c)(c \in d \Leftrightarrow c \in d)$ | UG 3-6 |
| 8. | $d = e$ | ACP |
| 9. | $(\forall c)(c \in d \Leftrightarrow c \in e)$ | IS 7, 8 |
| 10. | $d = e \Rightarrow (\forall c)(c \in d \Leftrightarrow c \in e)$ | CP 8-9 |
| 11. | $(\forall b)(d = b \Rightarrow (\forall c)(c \in d \Leftrightarrow c \in b))$ | UG 2-10 |
| 12. | $(\forall a)(\forall b)(a = b \Rightarrow (\forall c)(c \in a \Leftrightarrow c \in b))$ | UG 1-11 |

□

 
K. Suryanarayan, *Formal Proofs in Elementary Set Theory*, SpringerBriefs in Computational Intelligence, https://doi.org/10.1007/978-3-032-22239-8_2

**Axiom 2.1 (Zermelo Axiom: Axiom of Extensionality)** $(\forall a)(\forall b)((\forall c)(c \in a \Leftrightarrow c \in b) \Rightarrow a = b)$

**Axiom 2.2 (Zermelo Axiom: Axiom of Pairing)** $(\forall a)(\forall b)(\exists c)(\forall d)(d \in c \Leftrightarrow (d = a \vee d = b))$

**Theorem 2.2** $(\forall a)(\forall b)(\exists! c)(\forall d)(d \in c \Leftrightarrow (d = a \vee d = b))$

***Proof***

| | | |
|---|---|---|
| 1. | $(\forall a)(\forall b)(\exists c)(\forall d)(d \in c \Leftrightarrow (d = a \vee d = b))$ | Ax 2.2 |
| 2. | $(\forall a)(\forall b)((\forall c)(c \in a \Leftrightarrow c \in b) \Rightarrow a = b)$ | Ax 2.1 |
| 3. | flag $e$ | FSUG |
| 4. | $(\forall b)(\exists c)(\forall d)(d \in c \Leftrightarrow (d = e \vee d = b))$ | UI 1 |
| 5. | flag $f$ | FSUG |
| 6. | $(\exists c)(\forall d)(d \in c \Leftrightarrow (d = e \vee d = f))$ | UI 4 |
| 7. | $(\forall d)(d \in g \Leftrightarrow (d = e \vee d = f))$ | EI 6, flag $g$ |
| 8. | flag $h$ | FSUG |
| 9. | $(\forall d)(d \in h \Leftrightarrow (d = e \vee d = f))$ | ACP |
| 10. | flag $i$ | FSUG |
| 11. | $i \in g \Leftrightarrow (i = e \vee i = f)$ | UI 7 |
| 12. | $i \in h \Leftrightarrow (i = e \vee i = f)$ | UI 9 |
| 13. | $i \in g \Leftrightarrow i \in h$ | BT 11, 12 |
| 14. | $(\forall c)(c \in g \Leftrightarrow c \in h)$ | UG 10-13 |
| 15. | $(\forall b)((\forall c)(c \in g \Leftrightarrow c \in b) \Rightarrow g = b)$ | UI 2 |
| 16. | $(\forall c)(c \in g \Leftrightarrow c \in h) \Rightarrow g = h$ | UI 15 |
| 17. | $g = h$ | MP 14, 16 |
| 18. | $(\forall d)(d \in h \Leftrightarrow (d = e \vee d = f)) \Rightarrow g = h$ | CP 9-17 |
| 19. | $(\forall j)((\forall d)(d \in j \Leftrightarrow (d = e \vee d = f)) \Rightarrow g = j)$ | UG 8-18 |
| 20. | $(\forall d)(d \in g \Leftrightarrow (d = e \vee d = f)) \wedge$ $(\forall j)((\forall d)(d \in j \Leftrightarrow (d = e \vee d = f)) \Rightarrow g = j)$ | C 7, 19 |
| 21. | $(\exists c)((\forall d)(d \in c \Leftrightarrow (d = e \vee d = f)) \wedge$ $(\forall j)((\forall d)(d \in j \Leftrightarrow (d = e \vee d = f)) \Rightarrow c = j))$ | EG 20 |
| 22. | $(\exists! c)(\forall d)(d \in c \Leftrightarrow (d = e \vee d = f))$ | UQ 21 |
| 23. | $(\forall b)(\exists! c)(\forall d)(d \in c \Leftrightarrow (d = e \vee d = b))$ | UG 5-22 |
| 24. | $(\forall a)(\forall b)(\exists! c)(\forall d)(d \in c \Leftrightarrow (d = a \vee d = b))$ | UG 3-23 |

□

**Axiom 2.3 (Definition of Operation, from Theorem 2.2, Which If Applied to the Arguments $\alpha$ and $\beta$ Gives the Term "$\{\alpha, \beta\}$")**
$(\forall a)(\forall b)(\forall c)(c \in \{a, b\} \Leftrightarrow (c = a \vee c = b))$

**Theorem 2.3** $(\forall a)(\exists! b)(b = \{a, a\})$

***Proof***

| | | |
|---|---|---|
| 1. | flag $c$ | FSUG |
| 2. | $\{c, c\} = \{c, c\}$ | IR |
| 3. | $(\exists b)(b = \{c, c\})$ | EG 2 |
| 4. | flag $d$ | FSUG |
| 5. | flag $e$ | FSUG |
| 6. | $d = \{c, c\} \wedge e = \{c, c\}$ | ACP |
| 7. | $d = \{c, c\}$ | S 6 |
| 8. | $e = \{c, c\}$ | S 6 |
| 9. | $d = e$ | IT 7, 8 |
| 10. | $(d = \{c, c\} \wedge e = \{c, c\}) \Rightarrow d = e$ | CP 6-9 |
| 11. | $(\forall f)((d = \{c, c\} \wedge f = \{c, c\}) \Rightarrow d = f)$ | UG 5-10 |
| 12. | $(\forall g)(\forall f)((g = \{c, c\} \wedge f = \{c, c\}) \Rightarrow g = f)$ | UG 4-11 |
| 13. | $(\exists b)(b = \{c, c\}) \wedge$ $(\forall g)(\forall f)((g = \{c, c\} \wedge f = \{c, c\}) \Rightarrow g = f)$ | C 3, 12 |
| 14. | $(\exists! b)(b = \{c, c\})$ | UQ 13 |
| 15. | $(\forall a)(\exists! b)(b = \{a, a\})$ | UG 1-14 |

□

**Axiom 2.4 (Definition of Operation, from Theorem 2.3, Which If Applied to the Argument $\alpha$ Gives the Term "$\{\alpha\}$")**
$(\forall a)(\{a\} = \{a, a\})$

**Theorem 2.4** $(\forall a)(\forall b)(\{a, b\} = \{b, a\})$

***Proof***

| | | |
|---|---|---|
| 1. | $(\forall a)(\forall b)(\forall c)(c \in \{a, b\} \Leftrightarrow (c = a \vee c = b))$ | Ax 2.3 |
| 2. | $(\forall a)(\forall b)((\forall c)(c \in a \Leftrightarrow c \in b) \Rightarrow a = b)$ | Ax 2.1 |
| 3. | flag $d$ | FSUG |
| 4. | $(\forall b)(\forall c)(c \in \{d, b\} \Leftrightarrow (c = d \vee c = b))$ | UI 1 |
| 5. | flag $e$ | FSUG |
| 6. | $(\forall c)(c \in \{d, e\} \Leftrightarrow (c = d \vee c = e))$ | UI 4 |
| 7. | flag $f$ | FSUG |
| 8. | $f \in \{d, e\} \Leftrightarrow (f = d \vee f = e)$ | UI 6 |
| 9. | $f \in \{d, e\} \Leftrightarrow (f = e \vee f = d)$ | Co 8 |
| 10. | $(\forall b)(\forall c)(c \in \{e, b\} \Leftrightarrow (c = e \vee c = b))$ | UI 1 |
| 11. | $(\forall c)(c \in \{e, d\} \Leftrightarrow (c = e \vee c = d))$ | UI 10 |
| 12. | $f \in \{e, d\} \Leftrightarrow (f = e \vee f = d)$ | UI 11 |
| 13. | $f \in \{d, e\} \Leftrightarrow f \in \{e, d\}$ | BT 9, 12 |
| 14. | $(\forall c)(c \in \{d, e\} \Leftrightarrow c \in \{e, d\})$ | UG 7-13 |

| | | |
|---|---|---|
| 15. | $(\forall b)((\forall c)(c \in \{d, e\} \Leftrightarrow c \in b) \Rightarrow \{d, e\} = b)$ | UI 2 |
| 16. | $(\forall c)(c \in \{d, e\} \Leftrightarrow c \in \{e, d\}) \Rightarrow \{d, e\} = \{e, d\}$ | UI 15 |
| 17. | $\{d, e\} = \{e, d\}$ | MP 14, 16 |
| 18. | $(\forall b)(\{d, b\} = \{b, d\})$ | UG 5-17 |
| 19. | $(\forall a)(\forall b)(\{a, b\} = \{b, a\})$ | UG 3-18 |

□

**Theorem 2.5** $(\forall a)(\forall b)(a \in \{a, b\})$

***Proof***

| | | |
|---|---|---|
| 1. | $(\forall a)(\forall b)(\forall c)(c \in \{a, b\} \Leftrightarrow (c = a \vee c = b))$ | Ax 2.3 |
| 2. | flag $d$ | FSUG |
| 3. | $d = d$ | IR |
| 4. | flag $e$ | FSUG |
| 5. | $d = d \vee d = e$ | Ad 3 |
| 6. | $(\forall b)(\forall c)(c \in \{d, b\} \Leftrightarrow (c = d \vee c = b))$ | UI 1 |
| 7. | $(\forall c)(c \in \{d, e\} \Leftrightarrow (c = d \vee c = e))$ | UI 6 |
| 8. | $d \in \{d, e\} \Leftrightarrow (d = d \vee d = e)$ | UI 7 |
| 9. | $d \in \{d, e\}$ | BEl 5, 8 |
| 10. | $(\forall b)(d \in \{d, b\})$ | UG 4-9 |
| 11. | $(\forall a)(\forall b)(a \in \{a, b\})$ | UG 2-10 |

□

**Theorem 2.6** $(\forall a)(a \in \{a\})$

***Proof***

| | | |
|---|---|---|
| 1. | $(\forall a)(\forall b)(a \in \{a, b\})$ | T 2.5 |
| 2. | $(\forall a)(\{a\} = \{a, a\})$ | Ax 2.4 |
| 3. | flag $c$ | FSUG |
| 4. | $(\forall b)(c \in \{c, b\})$ | UI 1 |
| 5. | $c \in \{c, c\}$ | UI 4 |
| 6. | $\{c\} = \{c, c\}$ | UI 2 |
| 7. | $c \in \{c\}$ | IS 5, 6 |
| 8. | $(\forall a)(a \in \{a\})$ | UG 3-7 |

□

**Theorem 2.7** $(\forall a)(\forall b)(a \in \{b, a\})$

***Proof***

| | | |
|---|---|---|
| 1. | $(\forall a)(\forall b)(\{a, b\} = \{b, a\})$ | T 2.4 |
| 2. | $(\forall a)(\forall b)(a \in \{a, b\})$ | T 2.5 |
| 3. | flag $c$ | FSUG |
| 4. | $(\forall b)(\{c, b\} = \{b, c\})$ | UI 1 |
| 5. | flag $d$ | FSUG |
| 6. | $\{c, d\} = \{d, c\}$ | UI 4 |
| 7. | $(\forall b)(c \in \{c, b\})$ | UI 2 |
| 8. | $c \in \{c, d\}$ | UI 7 |
| 9. | $c \in \{d, c\}$ | IS 6, 8 |
| 10. | $(\forall b)(c \in \{b, c\})$ | UG 5-9 |
| 11. | $(\forall a)(\forall b)(a \in \{b, a\})$ | UG 3-10 |

□

**Theorem 2.8** $(\forall a)(\forall b)(b \in \{a\} \Leftrightarrow b = a)$

***Proof***

| | | |
|---|---|---|
| 1. | $(\forall a)(\forall b)(\forall c)(c \in \{a, b\} \Leftrightarrow (c = a \vee c = b))$ | Ax 2.3 |
| 2. | $(\forall a)(\{a\} = \{a, a\})$ | Ax 2.4 |
| 3. | flag $d$ | FSUG |
| 4. | $(\forall b)(\forall c)(c \in \{d, b\} \Leftrightarrow (c = d \vee c = b))$ | UI 1 |
| 5. | $(\forall c)(c \in \{d, d\} \Leftrightarrow (c = d \vee c = d))$ | UI 4 |
| 6. | flag $e$ | FSUG |
| 7. | $e \in \{d, d\} \Leftrightarrow (e = d \vee e = d)$ | UI 5 |
| 8. | $\{d\} = \{d, d\}$ | UI 2 |
| 9. | $e \in \{d\} \Leftrightarrow (e = d \vee e = d)$ | IS 7, 8 |
| 10. | $e \in \{d\} \Leftrightarrow e = d$ | Du 9 |
| 11. | $(\forall b)(b \in \{d\} \Leftrightarrow b = d)$ | UG 6-10 |
| 12. | $(\forall a)(\forall b)(b \in \{a\} \Leftrightarrow b = a)$ | UG 3-11 |

□

**Theorem 2.9** $(\forall a)(\forall b)(b \in \{a\} \Leftrightarrow a = b)$

***Proof***

| | | |
|---|---|---|
| 1. | $(\forall a)(\forall b)(b \in \{a\} \Leftrightarrow b = a)$ | T 2.8 |
| 2. | flag $c$ | FSUG |
| 3. | $(\forall b)(b \in \{c\} \Leftrightarrow b = c)$ | UI 1 |
| 4. | flag $d$ | FSUG |
| 5. | $d \in \{c\} \Leftrightarrow d = c$ | UI 3 |
| 6. | $d \in \{c\}$ | ACP |
| 7. | $d = c$ | BEl 5, 6 |
| 8. | $c = d$ | ISy 7 |
| 9. | $d \in \{c\} \Rightarrow c = d$ | CP 6-8 |
| 10. | $c = d$ | ACP |
| 11. | $d = c$ | ISy 10 |
| 12. | $d \in \{c\}$ | BEl 5, 11 |
| 13. | $c = d \Rightarrow d \in \{c\}$ | CP 10-12 |
| 14. | $(d \in \{c\} \Rightarrow c = d) \wedge (c = d \Rightarrow d \in \{c\})$ | C 9, 13 |
| 15. | $d \in \{c\} \Leftrightarrow c = d$ | BE 14 |
| 16. | $(\forall b)(b \in \{c\} \Leftrightarrow c = b)$ | UG 4-15 |
| 17. | $(\forall a)(\forall b)(b \in \{a\} \Leftrightarrow a = b)$ | UG 2-16 |

□

**Theorem 2.10** $(\forall a)(\forall b)(\{a\} = \{b\} \Rightarrow a = b)$

***Proof***

| | | |
|---|---|---|
| 1. | $(\forall a)(a \in \{a\})$ | T 2.6 |
| 2. | $(\forall a)(\forall b)(b \in \{a\} \Leftrightarrow b = a)$ | T 2.8 |
| 3. | flag $c$ | FSUG |
| 4. | flag $d$ | FSUG |
| 5. | $\{c\} = \{d\}$ | ACP |
| 6. | $c \in \{c\}$ | UI 1 |
| 7. | $c \in \{d\}$ | IS 5, 6 |
| 8. | $(\forall b)(b \in \{d\} \Leftrightarrow b = d)$ | UI 2 |
| 9. | $c \in \{d\} \Leftrightarrow c = d$ | UI 8 |
| 10. | $c = d$ | BEl 7, 9 |
| 11. | $\{c\} = \{d\} \Rightarrow c = d$ | CP 5-10 |
| 12. | $(\forall b)(\{c\} = \{b\} \Rightarrow c = b)$ | UG 4-11 |
| 13. | $(\forall a)(\forall b)(\{a\} = \{b\} \Rightarrow a = b)$ | UG 3-12 |

□

**Theorem 2.11** $(\forall a)(\forall b)(\forall c)(\{a\} = \{b, c\} \Rightarrow ((a = b \wedge a = c) \wedge b = c))$

***Proof***

| | | |
|---|---|---|
| 1. | $(\forall a)(\forall b)(a \in \{a, b\})$ | T 2.5 |
| 2. | $(\forall a)(\forall b)(b \in \{a\} \Leftrightarrow b = a)$ | T 2.8 |
| 3. | $(\forall a)(\forall b)(a \in \{b, a\})$ | T 2.7 |
| 4. | flag $d$ | FSUG |
| 5. | flag $e$ | FSUG |
| 6. | flag $f$ | FSUG |
| 7. | $\{d\} = \{e, f\}$ | ACP |
| 8. | $(\forall b)(e \in \{e, b\})$ | UI 1 |
| 9. | $e \in \{e, f\}$ | UI 8 |
| 10. | $e \in \{d\}$ | IS 7, 9 |
| 11. | $(\forall b)(b \in \{d\} \Leftrightarrow b = d)$ | UI 2 |
| 12. | $e \in \{d\} \Leftrightarrow e = d$ | UI 11 |
| 13. | $e = d$ | BEl 10, 12 |
| 14. | $(\forall b)(f \in \{b, f\})$ | UI 3 |
| 15. | $f \in \{e, f\}$ | UI 14 |
| 16. | $f \in \{d\}$ | IS 7, 15 |
| 17. | $f \in \{d\} \Leftrightarrow f = d$ | UI 11 |
| 18. | $f = d$ | BEl 16, 17 |
| 19. | $d = e$ | ISy 13 |
| 20. | $d = f$ | ISy 18 |
| 21. | $e = f$ | IT 19, 20 |
| 22. | $d = e \wedge d = f$ | C 19, 20 |
| 23. | $(d = e \wedge d = f) \wedge e = f$ | C 21, 22 |
| 24. | $\{d\} = \{e, f\} \Rightarrow ((d = e \wedge d = f) \wedge e = f)$ | CP 7-23 |
| 25. | $(\forall c)(\{d\} = \{e, c\} \Rightarrow ((d = e \wedge d = c) \wedge e = c))$ | UG 6-24 |
| 26. | $(\forall b)(\forall c)(\{d\} = \{b, c\} \Rightarrow ((d = b \wedge d = c) \wedge b = c))$ | UG 5-25 |
| 27. | $(\forall a)(\forall b)(\forall c)(\{a\} = \{b, c\} \Rightarrow ((a = b \wedge a = c) \wedge b = c))$ | UG 4-26 |

□

**Theorem 2.12** $(\forall a)(\forall b)(\forall c)(\{a\} = \{b, c\} \Rightarrow a = b)$

***Proof***

| | | |
|---|---|---|
| 1. | $(\forall a)(\forall b)(\forall c)(\{a\} = \{b, c\} \Rightarrow ((a = b \wedge a = c) \wedge b = c))$ | T 2.11 |
| 2. | flag $d$ | FSUG |
| 3. | $(\forall b)(\forall c)(\{d\} = \{b, c\} \Rightarrow ((d = b \wedge d = c) \wedge b = c))$ | UI 1 |
| 4. | flag $e$ | FSUG |
| 5. | $(\forall c)(\{d\} = \{e, c\} \Rightarrow ((d = e \wedge d = c) \wedge e = c))$ | UI 3 |
| 6. | flag $f$ | FSUG |
| 7. | $\{d\} = \{e, f\} \Rightarrow ((d = e \wedge d = f) \wedge e = f)$ | UI 5 |
| 8. | $\{d\} = \{e, f\}$ | ACP |
| 9. | $(d = e \wedge d = f) \wedge e = f$ | MP 7, 8 |
| 10. | $d = e \wedge d = f$ | S 9 |
| 11. | $d = e$ | S 10 |
| 12. | $\{d\} = \{e, f\} \Rightarrow d = e$ | CP 8-11 |
| 13. | $(\forall c)(\{d\} = \{e, c\} \Rightarrow d = e)$ | UG 6-12 |
| 14. | $(\forall b)(\forall c)(\{d\} = \{b, c\} \Rightarrow d = b)$ | UG 4-13 |
| 15. | $(\forall a)(\forall b)(\forall c)(\{a\} = \{b, c\} \Rightarrow a = b)$ | UG 2-14 |

□

**Theorem 2.13** $(\forall a)(\forall b)(\forall c)(\{a\} = \{b, c\} \Rightarrow a = c)$

***Proof***

| | | |
|---|---|---|
| 1. | $(\forall a)(\forall b)(\forall c)(\{a\} = \{b, c\} \Rightarrow ((a = b \wedge a = c) \wedge b = c))$ | T 2.11 |
| 2. | flag $d$ | FSUG |
| 3. | $(\forall b)(\forall c)(\{d\} = \{b, c\} \Rightarrow ((d = b \wedge d = c) \wedge b = c)$ | UI 1 |
| 4. | flag $e$ | FSUG |
| 5. | $(\forall c)(\{d\} = \{e, c\} \Rightarrow ((d = e \wedge d = c) \wedge e = c)$ | UI 3 |
| 6. | flag $f$ | FSUG |
| 7. | $\{d\} = \{e, f\} \Rightarrow ((d = e \wedge d = f) \wedge e = f)$ | UI 5 |
| 8. | $\{d\} = \{e, f\}$ | ACP |
| 9. | $(d = e \wedge d = f) \wedge e = f$ | MP 7, 8 |
| 10. | $d = e \wedge d = f$ | S 9 |
| 11. | $d = f$ | S 10 |
| 12. | $\{d\} = \{e, f\} \Rightarrow d = f$ | CP 8-11 |
| 13. | $(\forall c)(\{d\} = \{e, c\} \Rightarrow d = c)$ | UG 6-12 |
| 14. | $(\forall b)(\forall c)(\{d\} = \{b, c\} \Rightarrow d = c)$ | UG 4-13 |
| 15. | $(\forall a)(\forall b)(\forall c)(\{a\} = \{b, c\} \Rightarrow a = c)$ | UG 2-14 |

□

**Theorem 2.14** $(\forall a)(\forall b)(\forall c)(\{a\} = \{b, c\} \Rightarrow b = c)$

***Proof***

| | | |
|---|---|---|
| 1. | $(\forall a)(\forall b)(\forall c)(\{a\} = \{b, c\} \Rightarrow ((a = b \wedge a = c) \wedge b = c))$ | T 2.11 |
| 2. | flag $d$ | FSUG |
| 3. | $(\forall b)(\forall c)(\{d\} = \{b, c\} \Rightarrow ((d = b \wedge d = c) \wedge b = c)$ | UI 1 |
| 4. | flag $e$ | FSUG |
| 5. | $(\forall c)(\{d\} = \{e, c\} \Rightarrow ((d = e \wedge d = c) \wedge e = c)$ | UI 3 |
| 6. | flag $f$ | FSUG |
| 7. | $\{d\} = \{e, f\} \Rightarrow ((d = e \wedge d = f) \wedge e = f)$ | UI 5 |
| 8. | $\{d\} = \{e, f\}$ | ACP |
| 9. | $(d = e \wedge d = f) \wedge e = f$ | MP 7, 8 |
| 10. | $e = f$ | S 9 |
| 11. | $\{d\} = \{e, f\} \Rightarrow e = f$ | CP 8-10 |
| 12. | $(\forall c)(\{d\} = \{e, c\} \Rightarrow e = c)$ | UG 6-11 |
| 13. | $(\forall b)(\forall c)(\{d\} = \{b, c\} \Rightarrow b = c)$ | UG 4-12 |
| 14. | $(\forall a)(\forall b)(\forall c)(\{a\} = \{b, c\} \Rightarrow b = c)$ | UG 2-13 |

□

**Axiom 2.5 (Definition of Predicate Which If Applied to the Arguments $\alpha$ and $\beta$ Gives the Atomic Formula "$\alpha \neq \beta$")**
$(\forall a)(\forall b)(a \neq b \Leftrightarrow \neg(a = b))$

**Theorem 2.15** $(\forall a)(\forall b)(a = b \vee a \neq b)$

***Proof***

| | | |
|---|---|---|
| 1. | $(\forall a)(\forall b)(a \neq b \Leftrightarrow \neg(a = b))$ | Ax 2.5 |
| 2. | flag $c$ | FSUG |
| 3. | flag $d$ | FSUG |
| 4. | $c = d \vee \neg(c = d)$ | EM |
| 5. | $(\forall b)(c \neq b \Leftrightarrow \neg(c = b))$ | UI 1 |
| 6. | $c \neq d \Leftrightarrow \neg(c = d)$ | UI 5 |
| 7. | $(c \neq d \Rightarrow \neg(c = d)) \wedge ((\neg(c = d)) \Rightarrow c \neq d)$ | BE 6 |
| 8. | $(\neg(c = d)) \Rightarrow c \neq d$ | S 7 |
| 9. | $c = d \vee c \neq d$ | D 4, 8 |
| 10. | $(\forall b)(c = b \vee c \neq b)$ | UG 3-9 |
| 11. | $(\forall a)(\forall b)(a = b \vee a \neq b)$ | UG 2-10 |

□

## 2.2 Union

**Axiom 2.6 (Zermelo Axiom: Axiom of Union)** $(\forall a)(\exists b)(\forall c)(c \in b \Leftrightarrow (\exists d)(c \in d \wedge d \in a))$

**Theorem 2.16** $(\forall a)(\exists! b)(\forall c)(c \in b \Leftrightarrow (\exists d)(c \in d \wedge d \in a))$

*Proof*

| | | |
|---|---|---|
| 1. | $(\forall a)(\exists b)(\forall c)(c \in b \Leftrightarrow (\exists d)(c \in d \wedge d \in a))$ | Ax 2.6 |
| 2. | $(\forall a)(\forall b)((\forall c)(c \in a \Leftrightarrow c \in b) \Rightarrow a = b)$ | Ax 2.1 |
| 3. | flag $e$ | FSUG |
| 4. | $(\exists b)(\forall c)(c \in b \Leftrightarrow (\exists d)(c \in d \wedge d \in e))$ | UI 1 |
| 5. | $(\forall c)(c \in f \Leftrightarrow (\exists d)(c \in d \wedge d \in e))$ | EI 4, flag $f$ |
| 6. | flag $g$ | FSUG |
| 7. | $(\forall c)(c \in g \Leftrightarrow (\exists d)(c \in d \wedge d \in e))$ | ACP |
| 8. | flag $h$ | FSUG |
| 9. | $h \in g \Leftrightarrow (\exists d)(h \in d \wedge d \in e)$ | UI 7 |
| 10. | $h \in f \Leftrightarrow (\exists d)(h \in d \wedge d \in e)$ | UI 5 |
| 11. | $h \in g \Leftrightarrow h \in f$ | BT 9, 10 |
| 12. | $(\forall c)(c \in g \Leftrightarrow c \in f)$ | UG 8-11 |
| 13. | $(\forall b)((\forall c)(c \in g \Leftrightarrow c \in b) \Rightarrow g = b)$ | UI 2 |
| 14. | $(\forall c)(c \in g \Leftrightarrow c \in f) \Rightarrow g = f$ | UI 13 |
| 15. | $g = f$ | MP 12, 14 |
| 16. | $(\forall c)(c \in g \Leftrightarrow (\exists d)(c \in d \wedge d \in e)) \Rightarrow g = f$ | CP 7-15 |
| 17. | $(\forall i)((\forall c)(c \in i \Leftrightarrow (\exists d)(c \in d \wedge d \in e)) \Rightarrow i = f)$ | UG 6-16 |
| 18. | $(\forall c)(c \in f \Leftrightarrow (\exists d)(c \in d \wedge d \in e)) \wedge$ $(\forall i)((\forall c)(c \in i \Leftrightarrow (\exists d)(c \in d \wedge d \in e)) \Rightarrow i = f)$ | C 5, 17 |
| 19. | $(\exists b)((\forall c)(c \in b \Leftrightarrow (\exists d)(c \in d \wedge d \in e)) \wedge$ $(\forall i)((\forall c)(c \in i \Leftrightarrow (\exists d)(c \in d \wedge d \in e)) \Rightarrow i = b))$ | EG 18 |
| 20. | $(\exists! b)(\forall c)(c \in b \Leftrightarrow (\exists d)(c \in d \wedge d \in e))$ | UQ 19 |
| 21. | $(\forall a)(\exists! b)(\forall c)(c \in b \Leftrightarrow (\exists d)(c \in d \wedge d \in a))$ | UG 3-20 |

□

**Axiom 2.7 (Definitions of Operations, from Theorem 2.16, Which If Applied to the Argument $\alpha$ Give the Terms "$\cup\,\alpha$", "sum set of $\alpha$" and "union of elements of $\alpha$")**
$(\forall a)(\forall b)(b \in \cup a \Leftrightarrow (\exists c)(b \in c \wedge c \in a))$
$(\forall a)(\forall b)(b \in$ sum set of $a \Leftrightarrow (\exists c)(b \in c \wedge c \in a))$
$(\forall a)(\forall b)(b \in$ union of elements of $a \Leftrightarrow (\exists c)(b \in c \wedge c \in a))$

**Theorem 2.17** $(\forall a)(\forall b)(\exists! c)(c = \cup\{a, b\})$

***Proof***

| | | |
|---|---|---|
| 1. | flag $d$ | FSUG |
| 2. | flag $e$ | FSUG |
| 3. | $\cup\{d, e\} = \cup\{d, e\}$ | IR |
| 4. | $(\exists c)(c = \cup\{d, e\})$ | EG 3 |
| 5. | flag $f$ | FSUG |
| 6. | flag $g$ | FSUG |
| 7. | $f = \cup\{d, e\} \wedge g = \cup\{d, e\}$ | ACP |
| 8. | $f = \cup\{d, e\}$ | S 7 |
| 9. | $g = \cup\{d, e\}$ | S 7 |
| 10. | $f = g$ | IT 8, 9 |
| 11. | $(f = \cup\{d, e\} \wedge g = \cup\{d, e\}) \Rightarrow f = g$ | CP 7-10 |
| 12. | $(\forall i)((f = \cup\{d, e\} \wedge i = \cup\{d, e\}) \Rightarrow f = i)$ | UG 6-11 |
| 13. | $(\forall h)(\forall i)((h = \cup\{d, e\} \wedge i = \cup\{d, e\}) \Rightarrow h = i)$ | UG 5-12 |
| 14. | $(\exists c)(c = \cup\{d, e\}) \wedge$ $(\forall h)(\forall i)((h = \cup\{d, e\} \wedge i = \cup\{d, e\}) \Rightarrow h = i)$ | C 4, 13 |
| 15. | $(\exists! c)(c = \cup\{d, e\})$ | UQ 14 |
| 16. | $(\forall b)(\exists! c)(c = \cup\{d, b\})$ | UG 2-15 |
| 17. | $(\forall a)(\forall b)(\exists! c)(c = \cup\{a, b\})$ | UG 1-16 |

□

**Axiom 2.8 (Definitions of Operations, from Theorem 2.17, Which If Applied to the Arguments $\alpha$ and $\beta$ Give the Terms "$\alpha \cup \beta$" and "union of $\alpha$ and $\beta$")**
$(\forall a)(\forall b)(a \cup b = \cup\{a, b\})$
$(\forall a)(\forall b)(\text{union of } a \text{ and } b = \cup\{a, b\})$

**Theorem 2.18** $(\forall a)(\forall b)(\forall c)(c \in a \cup b \Leftrightarrow (c \in a \vee c \in b))$

***Proof***

| | | |
|---|---|---|
| 1. | $(\forall a)(\forall b)(b \in \cup a \Leftrightarrow (\exists c)(b \in c \wedge c \in a))$ | Ax 2.7 |
| 2. | $(\forall a)(\forall b)(a \cup b = \cup\{a, b\})$ | Ax 2.8 |
| 3. | $(\forall a)(\forall b)(\forall c)(c \in \{a, b\} \Leftrightarrow (c = a \vee c = b))$ | Ax 2.3 |
| 4. | $(\forall a)(\forall b)(a \in \{a, b\})$ | T 2.5 |
| 5. | $(\forall a)(\forall b)(a \in \{b, a\})$ | T 2.7 |
| 6. | flag $d$ | FSUG |
| 7. | flag $e$ | FSUG |
| 8. | $(\forall b)(b \in \cup\{d, e\} \Leftrightarrow (\exists c)(b \in c \wedge c \in \{d, e\}))$ | UI 1 |
| 9. | flag $f$ | FSUG |
| 10. | $f \in \cup\{d, e\} \Leftrightarrow (\exists c)(f \in c \wedge c \in \{d, e\})$ | UI 8 |

| | | |
|---|---|---|
| 11. | $(\forall b)(d \cup b = \cup\{d, b\})$ | UI 2 |
| 12. | $d \cup e = \cup\{d, e\}$ | UI 11 |
| 13. | $f \in d \cup e \Leftrightarrow (\exists c)(f \in c \wedge c \in \{d, e\})$ | IS 10, 12 |
| 14. | $\quad f \in d \cup e$ | ACP |
| 15. | $\quad (\exists c)(f \in c \wedge c \in \{d, e\})$ | BEl 13, 14 |
| 16. | $\quad f \in g \wedge g \in \{d, e\}$ | EI 15, flag $g$ |
| 17. | $\quad g \in \{d, e\}$ | S 16 |
| 18. | $\quad (\forall b)(\forall c)(c \in \{d, b\} \Leftrightarrow (c = d \vee c = b))$ | UI 3 |
| 19. | $\quad (\forall c)(c \in \{d, e\} \Leftrightarrow (c = d \vee c = e))$ | UI 18 |
| 20. | $\quad g \in \{d, e\} \Leftrightarrow (g = d \vee g = e)$ | UI 19 |
| 21. | $\quad g = d \vee g = e$ | BEl 17, 20 |
| 22. | $\quad f \in g$ | S 16 |
| 23. | $\quad f \in g \wedge (g = d \vee g = e)$ | C 21, 22 |
| 24. | $\quad (f \in g \wedge g = d) \vee (f \in g \wedge g = e)$ | Di 23 |
| 25. | $\qquad f \in g \wedge g = d$ | ACP |
| 26. | $\qquad f \in g$ | S 25 |
| 27. | $\qquad g = d$ | S 25 |
| 28. | $\qquad f \in d$ | IS 26, 27 |
| 29. | $\quad (f \in g \wedge g = d) \Rightarrow f \in d$ | CP 25-28 |
| 30. | $\qquad f \in g \wedge g = e$ | ACP |
| 31. | $\qquad f \in g$ | S 30 |
| 32. | $\qquad g = e$ | S 30 |
| 33. | $\qquad f \in e$ | IS 31, 32 |
| 34. | $\quad (f \in g \wedge g = e) \Rightarrow f \in e$ | CP 30-33 |
| 35. | $\quad f \in d \vee f \in e$ | D 24, 29, 34 |
| 36. | $f \in d \cup e \Rightarrow (f \in d \vee f \in e)$ | CP 14-35 |
| 37. | $\quad f \in d \vee f \in e$ | ACP |
| 38. | $\quad (\forall b)(d \in \{d, b\})$ | UI 4 |
| 39. | $\quad d \in \{d, e\}$ | UI 38 |
| 40. | $\quad (\forall b)(e \in \{b, e\})$ | UI 5 |
| 41. | $\quad e \in \{d, e\}$ | UI 40 |
| 42. | $\qquad \neg(\exists c)(f \in c \wedge c \in \{d, e\})$ | AIP |
| 43. | $\qquad (\forall c)\neg(f \in c \wedge c \in \{d, e\})$ | QN 42 |
| 44. | $\qquad \neg(f \in d \wedge d \in \{d, e\})$ | UI 43 |
| 45. | $\qquad (\neg(f \in d)) \vee \neg(d \in \{d, e\})$ | DM 44 |
| 46. | $\qquad \neg\neg(d \in \{d, e\})$ | DN 39 |
| 47. | $\qquad \neg(f \in d)$ | DS 45, 46 |
| 48. | $\qquad \neg(f \in e \wedge e \in \{d, e\})$ | UI 43 |
| 49. | $\qquad (\neg(f \in e)) \vee \neg(e \in \{d, e\})$ | DM 48 |
| 50. | $\qquad \neg\neg(e \in \{d, e\})$ | DN 41 |
| 51. | $\qquad \neg(f \in e)$ | DS 49, 50 |
| 52. | $\qquad (\neg(f \in d)) \wedge \neg(f \in e)$ | C 47, 51 |

| | | |
|---|---|---|
| 53. | $\neg(f \in d \vee f \in e)$ | DM 52 |
| 54. | $(f \in d \vee f \in e) \wedge \neg(f \in d \vee f \in e)$ | C 37, 53 |
| 55. | $\neg\neg(\exists c)(f \in c \wedge c \in \{d, e\})$ | IP 42-54 |
| 56. | $(\exists c)(f \in c \wedge c \in \{d, e\})$ | DN 55 |
| 57. | $f \in d \cup e$ | BEl 13, 56 |
| 58. | $(f \in d \vee f \in e) \Rightarrow f \in d \cup e$ | CP 37-57 |
| 59. | $(f \in d \cup e \Rightarrow (f \in d \vee f \in e)) \wedge$ $((f \in d \vee f \in e) \Rightarrow f \in d \cup e)$ | C 36, 58 |
| 60. | $f \in d \cup e \Leftrightarrow (f \in d \vee f \in e)$ | BE 59 |
| 61. | $(\forall c)(c \in d \cup e \Leftrightarrow (c \in d \vee c \in e))$ | UG 9-60 |
| 62. | $(\forall b)(\forall c)(c \in d \cup b \Leftrightarrow (c \in d \vee c \in b))$ | UG 7-61 |
| 63. | $(\forall a)(\forall b)(\forall c)(c \in a \cup b \Leftrightarrow (c \in a \vee c \in b))$ | UG 6-62 |

□

**Theorem 2.19** $(\forall a)(\forall b)(a \cup b = b \cup a)$

***Proof***

| | | |
|---|---|---|
| 1. | $(\forall a)(\forall b)(\forall c)(c \in a \cup b \Leftrightarrow (c \in a \vee c \in b))$ | T 2.18 |
| 2. | $(\forall a)(\forall b)((\forall c)(c \in a \Leftrightarrow c \in b) \Rightarrow a = b)$ | Ax 2.1 |
| 3. | flag $d$ | FSUG |
| 4. | $(\forall b)(\forall c)(c \in d \cup b \Leftrightarrow (c \in d \vee c \in b))$ | UI 1 |
| 5. | flag $e$ | FSUG |
| 6. | $(\forall c)(c \in d \cup e \Leftrightarrow (c \in d \vee c \in e))$ | UI 4 |
| 7. | flag $f$ | FSUG |
| 8. | $f \in d \cup e \Leftrightarrow (f \in d \vee f \in e)$ | UI 6 |
| 9. | $(\forall b)(\forall c)(c \in e \cup b \Leftrightarrow (c \in e \vee c \in b))$ | UI 1 |
| 10. | $(\forall c)(c \in e \cup d \Leftrightarrow (c \in e \vee c \in d))$ | UI 9 |
| 11. | $f \in e \cup d \Leftrightarrow (f \in e \vee f \in d)$ | UI 10 |
| 12. | $f \in e \cup d \Leftrightarrow (f \in d \vee f \in e)$ | Co 11 |
| 13. | $f \in d \cup e \Leftrightarrow f \in e \cup d$ | BT 8, 12 |
| 14. | $(\forall c)(c \in d \cup e \Leftrightarrow c \in e \cup d)$ | UG 7-13 |
| 15. | $(\forall b)((\forall c)(c \in d \cup e \Leftrightarrow c \in b) \Rightarrow d \cup e = b)$ | UI 2 |
| 16. | $(\forall c)(c \in d \cup e \Leftrightarrow c \in e \cup d) \Rightarrow d \cup e = e \cup d$ | UI 15 |
| 17. | $d \cup e = e \cup d$ | MP 14, 16 |
| 18. | $(\forall b)(d \cup b = b \cup d)$ | UG 5-17 |
| 19. | $(\forall a)(\forall b)(a \cup b = b \cup a)$ | UG 3-18 |

□

**Theorem 2.20** $(\forall a)(\forall b)(\forall c)((a \cup b) \cup c = a \cup (b \cup c))$

***Proof***

| | | |
|---|---|---|
| 1. | $(\forall a)(\forall b)(\forall c)(c \in a \cup b \Leftrightarrow (c \in a \vee c \in b))$ | T 2.18 |
| 2. | $(\forall a)(\forall b)((\forall c)(c \in a \Leftrightarrow c \in b) \Rightarrow a = b)$ | Ax 2.1 |
| 3. | flag $d$ | FSUG |
| 4. | flag $e$ | FSUG |
| 5. | $(\forall b)(\forall c)(c \in (d \cup e) \cup b \Leftrightarrow (c \in d \cup e \vee c \in b))$ | UI 1 |
| 6. | flag $f$ | FSUG |
| 7. | $(\forall c)(c \in (d \cup e) \cup f \Leftrightarrow (c \in d \cup e \vee c \in f))$ | UI 5 |
| 8. | flag $g$ | FSUG |
| 9. | $g \in (d \cup e) \cup f \Leftrightarrow (g \in d \cup e \vee g \in f)$ | UI 7 |
| 10. | $(\forall b)(\forall c)(c \in d \cup b \Leftrightarrow (c \in d \vee c \in b))$ | UI 1 |
| 11. | $(\forall c)(c \in d \cup e \Leftrightarrow (c \in d \vee c \in e))$ | UI 10 |
| 12. | $g \in d \cup e \Leftrightarrow (g \in d \vee g \in e)$ | UI 11 |
| 13. | $g \in d \cup e \vee g \in f$ | ACP |
| 14. | $(g \in d \vee g \in e) \vee g \in f$ | D 12, 13 |
| 15. | $(g \in d \cup e \vee g \in f) \Rightarrow ((g \in d \vee g \in e) \vee g \in f)$ | CP 13-14 |
| 16. | $(g \in d \vee g \in e) \vee g \in f$ | ACP |
| 17. | $g \in d \cup e \vee g \in f$ | D 12, 16 |
| 18. | $((g \in d \vee g \in e) \vee g \in f) \Rightarrow (g \in d \cup e \vee g \in f)$ | CP 16-17 |
| 19. | $((g \in d \cup e \vee g \in f) \Rightarrow ((g \in d \vee g \in e) \vee g \in f))\wedge$ $(((g \in d \vee g \in e) \vee g \in f) \Rightarrow (g \in d \cup e \vee g \in f))$ | C 15, 18 |
| 20. | $(g \in d \cup e \vee g \in f) \Leftrightarrow ((g \in d \vee g \in e) \vee g \in f)$ | BE 19 |
| 21. | $g \in (d \cup e) \cup f \Leftrightarrow ((g \in d \vee g \in e) \vee g \in f)$ | BT 9, 20 |
| 22. | $g \in (d \cup e) \cup f \Leftrightarrow (g \in d \vee (g \in e \vee g \in f))$ | A 21 |
| 23. | $(\forall c)(c \in d \cup (e \cup f) \Leftrightarrow (c \in d \vee c \in e \cup f))$ | UI 10 |
| 24. | $g \in d \cup (e \cup f) \Leftrightarrow (g \in d \vee g \in e \cup f)$ | UI 23 |
| 25. | $(\forall b)(\forall c)(c \in e \cup b \Leftrightarrow (c \in e \vee c \in b))$ | UI 1 |
| 26. | $(\forall c)(c \in e \cup f \Leftrightarrow (c \in e \vee c \in f))$ | UI 25 |
| 27. | $g \in e \cup f \Leftrightarrow (g \in e \vee g \in f)$ | UI 26 |
| 28. | $g \in d \vee g \in e \cup f$ | ACP |
| 29. | $g \in d \vee (g \in e \vee g \in f)$ | D 27, 28 |
| 30. | $(g \in d \vee g \in e \cup f) \Rightarrow (g \in d \vee (g \in e \vee g \in f))$ | CP 28-29 |
| 31. | $g \in d \vee (g \in e \vee g \in f)$ | ACP |
| 32. | $g \in d \vee g \in e \cup f$ | D 27, 31 |
| 33. | $(g \in d \vee (g \in e \vee g \in f)) \Rightarrow (g \in d \vee g \in e \cup f)$ | CP 31-32 |
| 34. | $((g \in d \vee g \in e \cup f) \Rightarrow (g \in d \vee (g \in e \vee g \in f)))$ $\wedge ((g \in d \vee (g \in e \vee g \in f)) \Rightarrow (g \in d \vee g \in e \cup f))$ | C 30, 32 |
| 35. | $(g \in d \vee g \in e \cup f) \Leftrightarrow (g \in d \vee (g \in e \vee g \in f))$ | BE 34 |
| 36. | $g \in d \cup (e \cup f) \Leftrightarrow (g \in d \vee (g \in e \vee g \in f))$ | BT 24, 35 |
| 37. | $g \in (d \cup e) \cup f \Leftrightarrow g \in d \cup (e \cup f)$ | BT 22, 36 |
| 38. | $(\forall c)(c \in (d \cup e) \cup f \Leftrightarrow c \in d \cup (e \cup f))$ | UG 8-37 |

| | | |
|---|---|---|
| 39. | $(\forall b)((\forall c)(c \in (d \cup e) \cup f \Leftrightarrow c \in b)$ $\Rightarrow (d \cup e) \cup f = b)$ | UI 2 |
| 40. | $(\forall c)(c \in (d \cup e) \cup f \Leftrightarrow c \in d \cup (e \cup f))$ $\Rightarrow (d \cup e) \cup f = d \cup (e \cup f)$ | UI 39 |
| 41. | $(d \cup e) \cup f = d \cup (e \cup f)$ | MP 38, 40 |
| 42. | $(\forall c)((d \cup e) \cup c = d \cup (e \cup c))$ | UG 6-41 |
| 43. | $(\forall b)(\forall c)((d \cup b) \cup c = d \cup (b \cup c))$ | UG 4-42 |
| 44. | $(\forall a)(\forall b)(\forall c)((a \cup b) \cup c = a \cup (b \cup c))$ | UG 3-43 |

□

**Theorem 2.21** $(\forall a)(a \cup a = a)$

***Proof***

| | | |
|---|---|---|
| 1. | $(\forall a)(\forall b)(\forall c)(c \in a \cup b \Leftrightarrow (c \in a \vee c \in b))$ | T 2.18 |
| 2. | $(\forall a)(\forall b)((\forall c)(c \in a \Leftrightarrow c \in b) \Rightarrow a = b)$ | Ax 2.1 |
| 3. | flag $d$ | FSUG |
| 4. | $(\forall b)(\forall c)(c \in d \cup b \Leftrightarrow (c \in d \vee c \in b))$ | UI 1 |
| 5. | $(\forall c)(c \in d \cup d \Leftrightarrow (c \in d \vee c \in d))$ | UI 4 |
| 6. | flag $e$ | FSUG |
| 7. | $e \in d \cup d \Leftrightarrow (e \in d \vee e \in d)$ | UI 5 |
| 8. | $e \in d \cup d \Leftrightarrow e \in d$ | Du 7 |
| 9. | $(\forall c)(c \in d \cup d \Leftrightarrow c \in d)$ | UG 6-8 |
| 10. | $(\forall b)((\forall c)(c \in d \cup d \Leftrightarrow c \in b) \Rightarrow d \cup d = b)$ | UI 2 |
| 11. | $(\forall c)(c \in d \cup d \Leftrightarrow c \in d) \Rightarrow d \cup d = d$ | UI 10 |
| 12. | $d \cup d = d$ | MP 9, 11 |
| 13. | $(\forall a)(a \cup a = a)$ | UG 3-12 |

□

**Theorem 2.22** $(\forall a)(\cup\{a\} = a)$

***Proof***

| | | |
|---|---|---|
| 1. | $(\forall a)(\{a\} = \{a, a\})$ | Ax 2.4 |
| 2. | $(\forall a)(\forall b)(a \cup b = \cup\{a, b\})$ | Ax 2.8 |
| 3. | $(\forall a)(a \cup a = a)$ | T 2.21 |
| 4. | flag $c$ | FSUG |
| 5. | $\{c\} = \{c, c\}$ | UI 1 |
| 6. | $(\forall b)(c \cup b = \cup\{c, b\})$ | UI 2 |
| 7. | $c \cup c = \cup\{c, c\}$ | UI 6 |
| 8. | $c \cup c = c$ | UI 3 |
| 9. | $\cup\{c, c\} = c$ | IT 7,8 |
| 10. | $\cup\{c\} = c$ | IS 5, 9 |
| 11. | $(\forall a)(\cup\{a\} = a)$ | UG 4-10 |

□

**Theorem 2.23** $(\forall a)(\forall b)(\{a\} \cup \{b\} = \{a, b\})$

***Proof***

| | | |
|---|---|---|
| 1. | $(\forall a)(\forall b)(\forall c)(c \in a \cup b \Leftrightarrow (c \in a \vee c \in b))$ | T 2.18 |
| 2. | $(\forall a)(\forall b)(b \in \{a\} \Leftrightarrow b = a)$ | T 2.8 |
| 3. | $(\forall a)(\forall b)(\forall c)(c \in \{a, b\} \Leftrightarrow (c = a \vee c = b))$ | Ax 2.3 |
| 4. | $(\forall a)(a \in \{a\})$ | T 2.6 |
| 5. | $(\forall a)(\forall b)((\forall c)(c \in a \Leftrightarrow c \in b) \Rightarrow a = b)$ | Ax 2.1 |
| 6. | flag $d$ | FSUG |
| 7. | flag $e$ | FSUG |
| 8. | flag $f$ | FSUG |
| 9. | $f \in \{d\} \cup \{e\}$ | ACP |
| 10. | $(\forall b)(\forall c)(c \in \{d\} \cup b \Leftrightarrow (c \in \{d\} \vee c \in b))$ | UI 1 |
| 11. | $(\forall c)(c \in \{d\} \cup \{e\} \Leftrightarrow (c \in \{d\} \vee c \in \{e\}))$ | UI 10 |
| 12. | $f \in \{d\} \cup \{e\} \Leftrightarrow (f \in \{d\} \vee f \in \{e\})$ | UI 11 |
| 13. | $f \in \{d\} \vee f \in \{e\}$ | BEl 9, 12 |
| 14. | $(\forall b)(b \in \{d\} \Leftrightarrow b = d)$ | UI 2 |
| 15. | $f \in \{d\} \Leftrightarrow f = d$ | UI 14 |
| 16. | $(\forall b)(b \in \{e\} \Leftrightarrow b = e)$ | UI 2 |
| 17. | $f \in \{e\} \Leftrightarrow f = e$ | UI 16 |
| 18. | $f = d \vee f = e$ | D 13, 15, 17 |
| 19. | $(\forall b)(\forall c)(c \in \{d, b\} \Leftrightarrow (c = d \vee c = b))$ | UI 3 |
| 20. | $(\forall c)(c \in \{d, e\} \Leftrightarrow (c = d \vee c = e))$ | UI 19 |
| 21. | $f \in \{d, e\} \Leftrightarrow (f = d \vee f = e)$ | UI 20 |
| 22. | $f \in \{d, e\}$ | BEl 18, 21 |
| 23. | $f \in \{d\} \cup \{e\} \Rightarrow f \in \{d, e\}$ | CP 9-22 |
| 24. | $f \in \{d, e\}$ | ACP |
| 25. | $(\forall b)(\forall c)(c \in \{d, b\} \Leftrightarrow (c = d \vee c = b))$ | UI 3 |
| 26. | $(\forall c)(c \in \{d, e\} \Leftrightarrow (c = d \vee c = e))$ | UI 25 |
| 27. | $f \in \{d, e\} \Leftrightarrow (f = d \vee f = e)$ | UI 26 |
| 28. | $f = d \vee f = e$ | BEl 24, 27 |
| 29. | $f = d$ | ACP |
| 30. | $d \in \{d\}$ | UI 4 |
| 31. | $f \in \{d\}$ | IS 29, 30 |
| 32. | $f = d \Rightarrow f \in \{d\}$ | CP 29-31 |
| 33. | $f = e$ | ACP |

| | | |
|---|---|---|
| 34. | $e \in \{e\}$ | UI 4 |
| 35. | $f \in \{e\}$ | IS 33, 34 |
| 36. | $f = e \Rightarrow f \in \{e\}$ | CP 33-35 |
| 37. | $f \in \{d\} \vee f \in \{e\}$ | D 28, 32, 36 |
| 38. | $(\forall b)(\forall c)(c \in \{d\} \cup b \Leftrightarrow (c \in \{d\} \vee c \in b))$ | UI 1 |
| 39. | $(\forall c)(c \in \{d\} \cup \{e\} \Leftrightarrow (c \in \{d\} \vee c \in \{e\}))$ | UI 38 |
| 40. | $f \in \{d\} \cup \{e\} \Leftrightarrow (f \in \{d\} \vee f \in \{e\})$ | UI 39 |
| 41. | $f \in \{d\} \cup \{e\}$ | BEI 37, 40 |
| 42. | $f \in \{d, e\} \Rightarrow f \in \{d\} \cup \{e\}$ | CP 24-41 |
| 43. | $(f \in \{d\} \cup \{e\} \Rightarrow f \in \{d, e\}) \wedge$ $(f \in \{d, e\} \Rightarrow f \in \{d\} \cup \{e\})$ | C 23, 42 |
| 44. | $f \in \{d\} \cup \{e\} \Leftrightarrow f \in \{d, e\}$ | BE 43 |
| 45. | $(\forall c)(c \in \{d\} \cup \{e\} \Leftrightarrow c \in \{d, e\})$ | UG 8-44 |
| 46. | $(\forall b)((\forall c)(c \in \{d\} \cup \{e\} \Leftrightarrow c \in b) \Rightarrow \{d\} \cup \{e\} = b)$ | UI 5 |
| 47. | $(\forall c)(c \in \{d\} \cup \{e\} \Leftrightarrow c \in \{d, e\}) \Rightarrow \{d\} \cup \{e\} = \{d, e\}$ | UI 46 |
| 48. | $\{d\} \cup \{e\} = \{d, e\}$ | MP 45, 47 |
| 49. | $(\forall b)(\{d\} \cup \{b\} = \{d, b\})$ | UG 7-48 |
| 50. | $(\forall a)(\forall b)(\{a\} \cup \{b\} = \{a, b\})$ | UG 6-49 |

□

## 2.3 Subset

**Axiom 2.9 (Definitions of Predicates Which If Applied to the Arguments $\alpha$ and $\beta$ Give the Atomic Formulas "$\alpha \subseteq \beta$" and "$\alpha$ is a subset of $\beta$")**

$(\forall a)(\forall b)(a \subseteq b \Leftrightarrow (\forall c)(c \in a \Rightarrow c \in b))$
$(\forall a)(\forall b)(a$ is a subset of $b \Leftrightarrow (\forall c)(c \in a \Rightarrow c \in b))$

**Theorem 2.24** $(\forall a)(\forall b)(a = b \Leftrightarrow (a \subseteq b \wedge b \subseteq a))$

***Proof***

| | | |
|---|---|---|
| 1. | $(\forall a)(\forall b)(a = b \Rightarrow (\forall c)(c \in a \Leftrightarrow c \in b))$ | T 2.1 |
| 2. | $(\forall a)(\forall b)(a \subseteq b \Leftrightarrow (\forall c)(c \in a \Rightarrow c \in b))$ | Ax 2.9 |
| 3. | $(\forall a)(\forall b)((\forall c)(c \in a \Leftrightarrow c \in b) \Rightarrow a = b)$ | Ax 2.1 |
| 4. | flag $d$ | FSUG |
| 5. | flag $e$ | FSUG |
| 6. | $d = e$ | ACP |
| 7. | $(\forall b)(d = b \Rightarrow (\forall c)(c \in d \Leftrightarrow c \in b))$ | UI 1 |
| 8. | $d = e \Rightarrow (\forall c)(c \in d \Leftrightarrow c \in e)$ | UI 7 |
| 9. | $(\forall c)(c \in d \Leftrightarrow c \in e)$ | MP 6, 8 |

| | | |
|---|---|---|
| 10. | flag $f$ | FSUG |
| 11. | $f \in d \Leftrightarrow f \in e$ | UI 9 |
| 12. | $(f \in d \Rightarrow f \in e) \wedge (f \in e \Rightarrow f \in d)$ | BE 11 |
| 13. | $f \in d \Rightarrow f \in e$ | S 12 |
| 14. | $(\forall c)(c \in d \Rightarrow c \in e)$ | UG 10-13 |
| 15. | $(\forall b)(d \subseteq b \Leftrightarrow (\forall c)(c \in d \Rightarrow c \in b))$ | UI 2 |
| 16. | $d \subseteq e \Leftrightarrow (\forall c)(c \in d \Rightarrow c \in e)$ | UI 15 |
| 17. | $d \subseteq e$ | BEI 14, 16 |
| 18. | flag $g$ | FSUG |
| 19. | $g \in d \Leftrightarrow g \in e$ | UI 9 |
| 20. | $(g \in d \Rightarrow g \in e) \wedge (g \in e \Rightarrow g \in d)$ | BE 19 |
| 21. | $g \in e \Rightarrow g \in d$ | S 20 |
| 22. | $(\forall c)(c \in e \Rightarrow c \in d)$ | UG 18-21 |
| 23. | $(\forall b)(e \subseteq b \Leftrightarrow (\forall c)(c \in e \Rightarrow c \in b))$ | UI 2 |
| 24. | $e \subseteq d \Leftrightarrow (\forall c)(c \in e \Rightarrow c \in d)$ | UI 23 |
| 25. | $e \subseteq d$ | BEI 22, 24 |
| 26. | $d \subseteq e \wedge e \subseteq d$ | C 17, 25 |
| 27. | $d = e \Rightarrow (d \subseteq e \wedge e \subseteq d)$ | CP 6-26 |
| 28. | $d \subseteq e \wedge e \subseteq d$ | ACP |
| 29. | $d \subseteq e$ | S 28 |
| 30. | $e \subseteq d$ | S 28 |
| 31. | $(\forall b)(d \subseteq b \Leftrightarrow (\forall c)(c \in d \Rightarrow c \in b))$ | UI 2 |
| 32. | $d \subseteq e \Leftrightarrow (\forall c)(c \in d \Rightarrow c \in e)$ | UI 31 |
| 33. | $(\forall c)(c \in d \Rightarrow c \in e)$ | BEI 29, 32 |
| 34. | $(\forall b)(e \subseteq b \Leftrightarrow (\forall c)(c \in e \Rightarrow c \in b))$ | UI 2 |
| 35. | $e \subseteq d \Leftrightarrow (\forall c)(c \in e \Rightarrow c \in d)$ | UI 34 |
| 36. | $(\forall c)(c \in e \Rightarrow c \in d)$ | BEI 30, 35 |
| 37. | flag $h$ | FSUG |
| 38. | $h \in d \Rightarrow h \in e$ | UI 33 |
| 39. | $h \in e \Rightarrow h \in d$ | UI 36 |
| 40. | $(h \in d \Rightarrow h \in e) \wedge (h \in e \Rightarrow h \in d)$ | C 38, 39 |
| 41. | $h \in d \Leftrightarrow h \in e$ | BE 40 |
| 42. | $(\forall c)(c \in d \Leftrightarrow c \in e)$ | UG 37-41 |
| 43. | $(\forall b)((\forall c)(c \in d \Leftrightarrow c \in b) \Rightarrow d = b)$ | UI 3 |
| 44. | $(\forall c)(c \in d \Leftrightarrow c \in e) \Rightarrow d = e$ | UI 43 |
| 45. | $d = e$ | MP 42, 44 |
| 46. | $(d \subseteq e \wedge e \subseteq d) \Rightarrow d = e$ | CP 28-45 |
| 47. | $(d = e \Rightarrow (d \subseteq e \wedge e \subseteq d)) \wedge ((d \subseteq e \wedge e \subseteq d) \Rightarrow d = e)$ | C 27, 46 |
| 48. | $d = e \Leftrightarrow (d \subseteq e \wedge e \subseteq d)$ | BE 47 |
| 49. | $(\forall b)(d = b \Leftrightarrow (d \subseteq b \wedge b \subseteq d))$ | UG 5-48 |
| 50. | $(\forall a)(\forall b)(a = b \Leftrightarrow (a \subseteq b \wedge b \subseteq a))$ | UG 4-49 |

□

**Theorem 2.25** $(\forall a)(a \subseteq a)$

***Proof***

| | | |
|---|---|---|
| 1. | $(\forall a)(\forall b)(a \subseteq b \Leftrightarrow (\forall c)(c \in a \Rightarrow c \in b))$ | Ax 2.9 |
| 2. | flag $d$ | FSUG |
| 3. | $(\forall b)(d \subseteq b \Leftrightarrow (\forall c)(c \in d \Rightarrow c \in b))$ | UI 1 |
| 4. | $d \subseteq d \Leftrightarrow (\forall c)(c \in d \Rightarrow c \in d)$ | UI 3 |
| 5. | flag $e$ | FSUG |
| 6. | $e \in d \Rightarrow e \in d$ | I |
| 7. | $(\forall c)(c \in d \Rightarrow c \in d)$ | UG 5-6 |
| 8. | $d \subseteq d$ | BEl 4, 7 |
| 9. | $(\forall a)(a \subseteq a)$ | UG 2-8 |

□

**Theorem 2.26** $(\forall a)(\forall b)(\forall c)((a \subseteq b \wedge b \subseteq c) \Rightarrow a \subseteq c)$

***Proof***

| | | |
|---|---|---|
| 1. | $(\forall a)(\forall b)(a \subseteq b \Leftrightarrow (\forall c)(c \in a \Rightarrow c \in b))$ | Ax 2.9 |
| 2. | flag $d$ | FSUG |
| 3. | flag $e$ | FSUG |
| 4. | flag $f$ | FSUG |
| 5. | $d \subseteq e \wedge e \subseteq f$ | ACP |
| 6. | $d \subseteq e$ | S 5 |
| 7. | $(\forall b)(d \subseteq b \Leftrightarrow (\forall c)(c \in d \Rightarrow c \in b))$ | UI 1 |
| 8. | $d \subseteq e \Leftrightarrow (\forall c)(c \in d \Rightarrow c \in e)$ | UI 7 |
| 9. | $(\forall c)(c \in d \Rightarrow c \in e)$ | BEl 6, 8 |
| 10. | $e \subseteq f$ | S 5 |
| 11. | $(\forall b)(e \subseteq b \Leftrightarrow (\forall c)(c \in e \Rightarrow c \in b))$ | UI 1 |
| 12. | $e \subseteq f \Leftrightarrow (\forall c)(c \in e \Rightarrow c \in f)$ | UI 11 |
| 13. | $(\forall c)(c \in e \Rightarrow c \in f)$ | BEl 10, 12 |
| 14. | flag $g$ | FSUG |
| 15. | $g \in d \Rightarrow g \in e$ | UI 9 |
| 16. | $g \in e \Rightarrow g \in f$ | UI 13 |
| 17. | $g \in d \Rightarrow g \in f$ | HS 15, 16 |
| 18. | $(\forall c)(c \in d \Rightarrow c \in f)$ | UG 14-17 |
| 19. | $d \subseteq f \Leftrightarrow (\forall c)(c \in d \Rightarrow c \in f)$ | UI 7 |
| 20. | $d \subseteq f$ | BEl 18, 19 |
| 21. | $(d \subseteq e \wedge e \subseteq f) \Rightarrow d \subseteq f$ | CP 5-20 |
| 22. | $(\forall c)((d \subseteq e \wedge e \subseteq c) \Rightarrow d \subseteq c)$ | UG 4-21 |
| 23. | $(\forall b)(\forall c)((d \subseteq b \wedge b \subseteq c) \Rightarrow d \subseteq c)$ | UG 3-22 |
| 24. | $(\forall a)(\forall b)(\forall c)((a \subseteq b \wedge b \subseteq c) \Rightarrow a \subseteq c)$ | UG 2-23 |

□

**Theorem 2.27** $(\forall a)(\forall b)(a \subseteq a \cup b)$

***Proof***

| | | |
|---|---|---|
| 1. | $(\forall a)(\forall b)(\forall c)(c \in a \cup b \Leftrightarrow (c \in a \vee c \in b))$ | T 2.18 |
| 2. | $(\forall a)(\forall b)(a \subseteq b \Leftrightarrow (\forall c)(c \in a \Rightarrow c \in b))$ | Ax 2.9 |
| 3. | flag $d$ | FSUG |
| 4. | flag $e$ | FSUG |
| 5. | flag $f$ | FSUG |
| 6. | $f \in d$ | ACP |
| 7. | $f \in d \vee f \in e$ | Ad 6 |
| 8. | $(\forall b)(\forall c)(c \in d \cup b \Leftrightarrow (c \in d \vee c \in b))$ | UI 1 |
| 9. | $(\forall c)(c \in d \cup e \Leftrightarrow (c \in d \vee c \in e))$ | UI 8 |
| 10. | $f \in d \cup e \Leftrightarrow (f \in d \vee f \in e)$ | UI 9 |
| 11. | $f \in d \cup e$ | BEl 7, 10 |
| 12. | $f \in d \Rightarrow f \in d \cup e$ | CP 6-11 |
| 13. | $(\forall c)(c \in d \Rightarrow c \in d \cup e)$ | UG 5-12 |
| 14. | $(\forall b)(d \subseteq b \Leftrightarrow (\forall c)(c \in d \Rightarrow c \in b))$ | UI 2 |
| 15. | $d \subseteq d \cup e \Leftrightarrow (\forall c)(c \in d \Rightarrow c \in d \cup e)$ | UI 14 |
| 16. | $d \subseteq d \cup e$ | BEl 13, 15 |
| 17. | $(\forall b)(d \subseteq d \cup b)$ | UG 4-16 |
| 18. | $(\forall a)(\forall b)(a \subseteq a \cup b)$ | UG 3-17 |

□

**Theorem 2.28** $(\forall a)(\forall b)(\forall c)(c \in a \Rightarrow c \in a \cup b)$

***Proof***

| | | |
|---|---|---|
| 1. | $(\forall a)(\forall b)(a \subseteq a \cup b)$ | T 2.27 |
| 2. | $(\forall a)(\forall b)(a \subseteq b \Leftrightarrow (\forall c)(c \in a \Rightarrow c \in b))$ | Ax 2.9 |
| 3. | flag $d$ | FSUG |
| 4. | flag $e$ | FSUG |
| 5. | $(\forall b)(d \subseteq d \cup b)$ | UI 1 |
| 6. | $d \subseteq d \cup e$ | UI 5 |
| 7. | $(\forall b)(d \subseteq b \Leftrightarrow (\forall c)(c \in d \Rightarrow c \in b))$ | UI 2 |
| 8. | $d \subseteq d \cup e \Leftrightarrow (\forall c)(c \in d \Rightarrow c \in d \cup e)$ | UI 7 |
| 9. | $(\forall c)(c \in d \Rightarrow c \in d \cup e)$ | BEl 6, 8 |
| 10. | $(\forall b)(\forall c)(c \in d \Rightarrow c \in d \cup b)$ | UG 4-9 |
| 11. | $(\forall a)(\forall b)(\forall c)(c \in a \Rightarrow c \in a \cup b)$ | UG 3-10 |

□

**Theorem 2.29** $(\forall a)(\forall b)(b \subseteq a \cup b)$

***Proof***

| | | |
|---|---|---|
| 1. | $(\forall a)(\forall b)(a \subseteq a \cup b)$ | T 2.27 |
| 2. | $(\forall a)(\forall b)(a \cup b = b \cup a)$ | T 2.19 |
| 3. | flag $c$ | FSUG |
| 4. | flag $d$ | FSUG |
| 5. | $(\forall b)(d \subseteq d \cup b)$ | UI 1 |
| 6. | $d \subseteq d \cup c$ | UI 5 |
| 7. | $(\forall b)(c \cup b = b \cup c)$ | UI 2 |
| 8. | $c \cup d = d \cup c$ | UI 7 |
| 9. | $d \subseteq c \cup d$ | IS 6, 8 |
| 10. | $(\forall b)(b \subseteq c \cup b)$ | UG 4-9 |
| 11. | $(\forall a)(\forall b)(b \subseteq a \cup b)$ | UG 3-10 |

□

**Theorem 2.30** $(\forall a)(\forall b)(\forall c)(c \in b \Rightarrow c \in a \cup b)$

***Proof***

| | | |
|---|---|---|
| 1. | $(\forall a)(\forall b)(b \subseteq a \cup b)$ | T 2.29 |
| 2. | $(\forall a)(\forall b)(a \subseteq b \Leftrightarrow (\forall c)(c \in a \Rightarrow c \in b))$ | Ax 2.9 |
| 3. | flag $d$ | FSUG |
| 4. | flag $e$ | FSUG |
| 5. | $(\forall b)(b \subseteq d \cup b)$ | UI 1 |
| 6. | $e \subseteq d \cup e$ | UI 5 |
| 7. | $(\forall b)(e \subseteq b \Leftrightarrow (\forall c)(c \in e \Rightarrow c \in b))$ | UI 2 |
| 8. | $e \subseteq d \cup e \Leftrightarrow (\forall c)(c \in e \Rightarrow c \in d \cup e)$ | UI 7 |
| 9. | $(\forall c)(c \in e \Rightarrow c \in d \cup e)$ | BEl 6, 8 |
| 10. | $(\forall b)(\forall c)(c \in b \Rightarrow c \in d \cup b)$ | UG 4-9 |
| 11. | $(\forall a)(\forall b)(\forall c)(c \in b \Rightarrow c \in a \cup b)$ | UG 3-10 |

□

**Theorem 2.31** $(\forall a)(\forall b)(\forall c)((a \in c \wedge b \in c) \Leftrightarrow \{a, b\} \subseteq c)$

***Proof***

| | | |
|---|---|---|
| 1. | $(\forall a)(\forall b)(\forall c)(c \in \{a, b\} \Leftrightarrow (c = a \vee c = b))$ | Ax 2.3 |
| 2. | $(\forall a)(\forall b)(a \subseteq b \Leftrightarrow (\forall c)(c \in a \Rightarrow c \in b))$ | Ax 2.9 |
| 3. | $(\forall a)(\forall b)(a \in \{a, b\})$ | T 2.5 |
| 4. | $(\forall a)(\forall b)(a \in \{b, a\})$ | T 2.7 |

| | | |
|---|---|---|
| 5. | flag $d$ | FSUG |
| 6. | flag $e$ | FSUG |
| 7. | flag $f$ | FSUG |
| 8. | $d \in f \wedge e \in f$ | ACP |
| 9. | $d \in f$ | S 8 |
| 10. | $e \in f$ | S 8 |
| 11. | flag $g$ | FSUG |
| 12. | $g \in \{d, e\}$ | ACP |
| 13. | $(\forall b)(\forall c)(c \in \{d, b\} \Leftrightarrow (c = d \vee c = b))$ | UI 1 |
| 14. | $(\forall c)(c \in \{d, e\} \Leftrightarrow (c = d \vee c = e))$ | UI 13 |
| 15. | $g \in \{d, e\} \Leftrightarrow (g = d \vee g = e)$ | UI 14 |
| 16. | $g = d \vee g = e$ | BEl 12, 15 |
| 17. | $g = d$ | ACP |
| 18. | $g \in f$ | IS 9, 17 |
| 19. | $g = d \Rightarrow g \in f$ | CP 17-18 |
| 20. | $g = e$ | ACP |
| 21. | $g \in f$ | IS 10, 20 |
| 22. | $g = e \Rightarrow g \in f$ | CP 20-21 |
| 23. | $g \in f \vee g \in f$ | D 16, 19, 22 |
| 24. | $g \in f$ | Du 23 |
| 25. | $g \in \{d, e\} \Rightarrow g \in f$ | CP 12-24 |
| 26. | $(\forall c)(c \in \{d, e\} \Rightarrow c \in f)$ | UG 11-25 |
| 27. | $(\forall b)(\{d, e\} \subseteq b \Leftrightarrow (\forall c)(c \in \{d, e\} \Rightarrow c \in b))$ | UI 2 |
| 28. | $\{d, e\} \subseteq f \Leftrightarrow (\forall c)(c \in \{d, e\} \Rightarrow c \in f)$ | UI 27 |
| 29. | $\{d, e\} \subseteq f$ | BEl 26, 28 |
| 30. | $(d \in f \wedge e \in f) \Rightarrow \{d, e\} \subseteq f$ | CP 8-29 |
| 31. | $\{d, e\} \subseteq f$ | ACP |
| 32. | $(\forall b)(\{d, e\} \subseteq b \Leftrightarrow (\forall c)(c \in \{d, e\} \Rightarrow c \in b))$ | UI 2 |
| 33. | $\{d, e\} \subseteq f \Leftrightarrow (\forall c)(c \in \{d, e\} \Rightarrow c \in f)$ | UI 32 |
| 34. | $(\forall c)(c \in \{d, e\} \Rightarrow c \in f)$ | BEl 31, 33 |
| 35. | $d \in \{d, e\} \Rightarrow d \in f$ | UI 34 |
| 36. | $(\forall b)(d \in \{d, b\})$ | UI 3 |
| 37. | $d \in \{d, e\}$ | UI 36 |
| 38. | $d \in f$ | MP 35, 37 |
| 39. | $e \in \{d, e\} \Rightarrow e \in f$ | UI 34 |
| 40. | $(\forall b)(e \in \{b, e\})$ | UI 4 |
| 41. | $e \in \{d, e\}$ | UI 40 |
| 42. | $e \in f$ | MP 39, 41 |
| 43. | $d \in f \wedge e \in f$ | C 38, 42 |
| 44. | $\{d, e\} \subseteq f \Rightarrow (d \in f \wedge e \in f)$ | CP 31-43 |
| 45. | $((d \in f \wedge e \in f) \Rightarrow \{d, e\} \subseteq f) \wedge$ $(\{d, e\} \subseteq f \Rightarrow (d \in f \wedge e \in f))$ | C 30, 44 |

| | | |
|---|---|---|
| 46. | $(d \in f \wedge e \in f) \Leftrightarrow \{d, e\} \subseteq f$ | BE 45 |
| 47. | $(\forall c)((d \in c \wedge e \in c) \Leftrightarrow \{d, e\} \subseteq c)$ | UG 7-46 |
| 48. | $(\forall b)(\forall c)((d \in c \wedge b \in c) \Leftrightarrow \{d, b\} \subseteq c)$ | UG 6-47 |
| 49. | $(\forall a)(\forall b)(\forall c)((a \in c \wedge b \in c) \Leftrightarrow \{a, b\} \subseteq c)$ | UG 5-48 |

□

**Theorem 2.32** $(\forall a)(\forall b)(a \in b \Leftrightarrow \{a\} \subseteq b)$

***Proof***

| | | |
|---|---|---|
| 1. | $(\forall a)(\forall b)(\forall c)((a \in c \wedge b \in c) \Leftrightarrow \{a, b\} \subseteq c)$ | T 2.31 |
| 2. | $(\forall a)(\{a\} = \{a, a\})$ | Ax 2.4 |
| 3. | flag $d$ | FSUG |
| 4. | flag $e$ | FSUG |
| 5. | $(\forall b)(\forall c)((d \in c \wedge b \in c) \Leftrightarrow \{d, b\} \subseteq c)$ | UI 1 |
| 6. | $(\forall c)((d \in c \wedge d \in c) \Leftrightarrow \{d, d\} \subseteq c)$ | UI 5 |
| 7. | $(d \in e \wedge d \in e) \Leftrightarrow \{d, d\} \subseteq e$ | UI 6 |
| 8. | $d \in e \Leftrightarrow \{d, d\} \subseteq e$ | Du 7 |
| 9. | $\{d\} = \{d, d\}$ | UI 2 |
| 10. | $d \in e \Leftrightarrow \{d\} \subseteq e$ | IS 8, 9 |
| 11. | $(\forall b)(d \in b \Leftrightarrow \{d\} \subseteq b)$ | UG 4-10 |
| 12. | $(\forall a)(\forall b)(a \in b \Leftrightarrow \{a\} \subseteq b)$ | UG 3-11 |

□

**Theorem 2.33** $(\forall a)(\forall b)(\forall c)(\forall d)((a \in b \wedge c \in d) \Rightarrow \{a, c\} \subseteq b \cup d)$

***Proof***

| | | |
|---|---|---|
| 1. | $(\forall a)(\forall b)(a \subseteq a \cup b)$ | T 2.27 |
| 2. | $(\forall a)(\forall b)(a \subseteq b \Leftrightarrow (\forall c)(c \in a \Rightarrow c \in b))$ | Ax 2.9 |
| 3. | $(\forall a)(\forall b)(b \subseteq a \cup b)$ | T 2.29 |
| 4. | $(\forall a)(\forall b)(\forall c)((a \in c \wedge b \in c) \Leftrightarrow \{a, b\} \subseteq c)$ | T 2.31 |
| 5. | flag $e$ | FSUG |
| 6. | flag $f$ | FSUG |
| 7. | flag $g$ | FSUG |
| 8. | flag $h$ | FSUG |
| 9. | $e \in f \wedge g \in h$ | ACP |
| 10. | $e \in f$ | S 9 |
| 11. | $(\forall b)(f \subseteq f \cup b)$ | UI 1 |
| 12. | $f \subseteq f \cup h$ | UI 11 |
| 13. | $(\forall b)(f \subseteq b \Leftrightarrow (\forall c)(c \in f \Rightarrow c \in b))$ | UI 2 |

| | | |
|---|---|---|
| 14. | $f \subseteq f \cup h \Leftrightarrow (\forall c)(c \in f \Rightarrow c \in f \cup h)$ | UI 13 |
| 15. | $(\forall c)(c \in f \Rightarrow c \in f \cup h)$ | BEl 12, 14 |
| 16. | $e \in f \Rightarrow e \in f \cup h$ | UI 15 |
| 17. | $e \in f \cup h$ | MP 10, 16 |
| 18. | $g \in h$ | S 9 |
| 19. | $(\forall b)(b \subseteq f \cup b)$ | UI 3 |
| 20. | $h \subseteq f \cup h$ | UI 19 |
| 21. | $(\forall b)(h \subseteq b \Leftrightarrow (\forall c)(c \in h \Rightarrow c \in b))$ | UI 2 |
| 22. | $h \subseteq f \cup h \Leftrightarrow (\forall c)(c \in h \Rightarrow c \in f \cup h)$ | UI 21 |
| 23. | $(\forall c)(c \in h \Rightarrow c \in f \cup h)$ | BEl 20, 22 |
| 24. | $g \in h \Rightarrow g \in f \cup h$ | UI 23 |
| 25. | $g \in f \cup h$ | MP 18, 24 |
| 26. | $e \in f \cup h \wedge g \in f \cup h$ | C 17, 25 |
| 27. | $(\forall b)(\forall c)((e \in c \wedge b \in c) \Leftrightarrow \{e, b\} \subseteq c)$ | UI 4 |
| 28. | $(\forall c)((e \in c \wedge g \in c) \Leftrightarrow \{e, g\} \subseteq c)$ | UI 27 |
| 29. | $(e \in f \cup h \wedge g \in f \cup h) \Leftrightarrow \{e, g\} \subseteq f \cup h$ | UI 28 |
| 30. | $\{e, g\} \subseteq f \cup h$ | BEl 26, 29 |
| 31. | $(e \in f \wedge g \in h) \Rightarrow \{e, g\} \subseteq f \cup h$ | CP 9-30 |
| 32. | $(\forall d)((e \in f \wedge g \in d) \Rightarrow \{e, g\} \subseteq f \cup d)$ | UG 8-31 |
| 33. | $(\forall c)(\forall d)((e \in f \wedge c \in d) \Rightarrow \{e, c\} \subseteq f \cup d)$ | UG 7-32 |
| 34. | $(\forall b)(\forall c)(\forall d)((e \in b \wedge c \in d) \Rightarrow \{e, c\} \subseteq b \cup d)$ | UG 6-33 |
| 35. | $(\forall a)(\forall b)(\forall c)(\forall d)((a \in b \wedge c \in d) \Rightarrow \{a, c\} \subseteq b \cup d)$ | UG 5-34 |

□

**Axiom 2.10 (Definitions of Predicates Which If Applied to the Arguments $\alpha$ and $\beta$ Give the Atomic Formulas "$\alpha \subsetneq \beta$" and "$\alpha$ is a proper subset of $\beta$")**

$(\forall a)(\forall b)(a \subsetneq b \Leftrightarrow (a \subseteq b \wedge a \neq b))$
$(\forall a)(\forall b)(a$ is a proper subset of $b \Leftrightarrow (a \subseteq b \wedge a \neq b))$

**Axiom 2.11 (Definitions of Predicates Which If Applied to the Arguments $\alpha$ and $\beta$ Give the Atomic Formulas "$\alpha \nsubseteq \beta$" and "$\alpha$ is not a subset of $\beta$")**

$(\forall a)(\forall b)(a \nsubseteq b \Leftrightarrow \neg(a \subseteq b))$
$(\forall a)(\forall b)(a$ is not a subset of $b \Leftrightarrow \neg(a \subseteq b))$

**Theorem 2.34** $(\forall a)(\forall b)(a \subsetneq b \Rightarrow b \nsubseteq a)$

***Proof***

| | | |
|---|---|---|
| 1. | $(\forall a)(\forall b)(a \subsetneq b \Leftrightarrow (a \subseteq b \wedge a \neq b))$ | Ax 2.10 |
| 2. | $(\forall a)(\forall b)(a \neq b \Leftrightarrow \neg(a = b))$ | Ax 2.5 |
| 3. | $(\forall a)(\forall b)(a = b \Leftrightarrow (a \subseteq b \wedge b \subseteq a))$ | T 2.24 |

| | | |
|---|---|---|
| 4. | $(\forall a)(\forall b)(a \not\subseteq b \Leftrightarrow \neg(a \subseteq b))$ | Ax 2.11 |
| 5. | flag $c$ | FSUG |
| 6. | $(\forall b)(c \subsetneq b \Leftrightarrow (c \subseteq b \wedge c \neq b))$ | UI 1 |
| 7. | flag $d$ | FSUG |
| 8. | $c \subsetneq d \Leftrightarrow (c \subseteq d \wedge c \neq d)$ | UI 6 |
| 9. | $c \subsetneq d$ | ACP |
| 10. | $c \subseteq d \wedge c \neq d$ | BEI 8, 9 |
| 11. | $c \subseteq d$ | S 10 |
| 12. | $c \neq d$ | S 10 |
| 13. | $(\forall b)(c \neq b \Leftrightarrow \neg(c = b))$ | UI 2 |
| 14. | $c \neq d \Leftrightarrow \neg(c = d)$ | UI 13 |
| 15. | $\neg(c = d)$ | BEI 12, 14 |
| 16. | $d \subseteq c$ | AIP |
| 17. | $c \subseteq d \wedge d \subseteq c$ | C 11, 16 |
| 18. | $(\forall b)(c = b \Leftrightarrow (c \subseteq b \wedge b \subseteq c))$ | UI 3 |
| 19. | $c = d \Leftrightarrow (c \subseteq d \wedge d \subseteq c)$ | UI 18 |
| 20. | $c = d$ | BEI 17, 19 |
| 21. | $c = d \wedge \neg(c = d)$ | C 15, 20 |
| 22. | $\neg(d \subseteq c)$ | IP 16-21 |
| 23. | $(\forall b)(d \not\subseteq b \Leftrightarrow \neg(d \subseteq b))$ | UI 4 |
| 24. | $d \not\subseteq c \Leftrightarrow \neg(d \subseteq c)$ | UI 23 |
| 25. | $d \not\subseteq c$ | BEI 22, 24 |
| 26. | $c \subsetneq d \Rightarrow d \not\subseteq c$ | CP 9-25 |
| 27. | $(\forall b)(c \subsetneq b \Rightarrow b \not\subseteq c)$ | UG 7-26 |
| 28. | $(\forall a)(\forall b)(a \subsetneq b \Rightarrow b \not\subseteq a)$ | UG 5-27 |

□

**Theorem 2.35** $(\forall a)(\forall b)(\forall c)((a \subsetneq b \wedge b \subseteq c) \Rightarrow a \subsetneq c)$

***Proof***

| | | |
|---|---|---|
| 1. | $(\forall a)(\forall b)(a \subsetneq b \Leftrightarrow (a \subseteq b \wedge a \neq b))$ | Ax 2.10 |
| 2. | $(\forall a)(\forall b)(a = b \Leftrightarrow (a \subseteq b \wedge b \subseteq a))$ | T 2.24 |
| 3. | $(\forall a)(\forall b)(\forall c)((a \subseteq b \wedge b \subseteq c) \Rightarrow a \subseteq c)$ | T 2.26 |
| 4. | $(\forall a)(\forall b)(a \neq b \Leftrightarrow \neg(a = b))$ | Ax 2.5 |
| 5. | flag $d$ | FSUG |
| 6. | flag $e$ | FSUG |
| 7. | flag $f$ | FSUG |
| 8. | $d \subsetneq e \wedge e \subseteq f$ | ACP |
| 9. | $d \subsetneq e$ | S 8 |
| 10. | $e \subseteq f$ | S 8 |
| 11. | $(\forall b)(d \subsetneq b \Leftrightarrow (d \subseteq b \wedge d \neq b))$ | UI 1 |

| | | |
|---|---|---|
| 12. | $d \subsetneq e \Leftrightarrow (d \subseteq e \wedge d \neq e)$ | UI 11 |
| 13. | $d \subseteq e \wedge d \neq e$ | BEl 9, 12 |
| 14. | $d \subseteq e$ | S 13 |
| 15. | $d \neq e$ | S 13 |
| 16. | $d = f$ | AIP |
| 17. | $(\forall b)(d = b \Leftrightarrow (d \subseteq b \wedge b \subseteq d))$ | UI 2 |
| 18. | $d = f \Leftrightarrow (d \subseteq f \wedge f \subseteq d)$ | UI 17 |
| 19. | $d \subseteq f \wedge f \subseteq d$ | BEl 16, 18 |
| 20. | $f \subseteq d$ | S 19 |
| 21. | $e \subseteq f \wedge f \subseteq d$ | C 10, 20 |
| 22. | $(\forall b)(\forall c)((e \subseteq b \wedge b \subseteq c) \Rightarrow e \subseteq c)$ | UI 3 |
| 23. | $(\forall c)((e \subseteq f \wedge f \subseteq c) \Rightarrow e \subseteq c)$ | UI 22 |
| 24. | $(e \subseteq f \wedge f \subseteq d) \Rightarrow e \subseteq d$ | UI 23 |
| 25. | $e \subseteq d$ | MP 21, 24 |
| 26. | $d \subseteq e \wedge e \subseteq d$ | C 14, 25 |
| 27. | $d = e \Leftrightarrow (d \subseteq e \wedge e \subseteq d)$ | UI 17 |
| 28. | $d = e$ | BEl 26, 27 |
| 29. | $(\forall b)(d \neq b \Leftrightarrow \neg(d = b))$ | UI 4 |
| 30. | $d \neq e \Leftrightarrow \neg(d = e)$ | UI 29 |
| 31. | $\neg(d = e)$ | BEl 15, 30 |
| 32. | $d = e \wedge \neg(d = e)$ | C 28, 31 |
| 33. | $\neg(d = f)$ | IP 16-32 |
| 34. | $(\forall b)(d \neq b \Leftrightarrow \neg(d = b))$ | UI 4 |
| 35. | $d \neq f \Leftrightarrow \neg(d = f))$ | UI 34 |
| 36. | $d \neq f$ | BEl 33, 35 |
| 37. | $d \subseteq e \wedge e \subseteq f$ | C 10, 14 |
| 38. | $(\forall b)(\forall c)((d \subseteq b \wedge b \subseteq c) \Rightarrow d \subseteq c)$ | UI 3 |
| 39. | $(\forall c)((d \subseteq e \wedge e \subseteq c) \Rightarrow d \subseteq c)$ | UI 38 |
| 40. | $(d \subseteq e \wedge e \subseteq f) \Rightarrow d \subseteq f$ | UI 39 |
| 41. | $d \subseteq f$ | MP 37, 40 |
| 42. | $d \subseteq f \wedge d \neq f$ | C 36, 41 |
| 43. | $d \subsetneq f \Leftrightarrow (d \subseteq f \wedge d \neq f)$ | UI 11 |
| 44. | $d \subsetneq f$ | BEl 42, 43 |
| 45. | $(d \subsetneq e \wedge e \subseteq f) \Rightarrow d \subsetneq f$ | CP 8-44 |
| 46. | $(\forall c)((d \subsetneq e \wedge e \subseteq c) \Rightarrow d \subsetneq c)$ | UG 7-45 |
| 47. | $(\forall b)(\forall c)((d \subsetneq b \wedge b \subseteq c) \Rightarrow d \subsetneq c)$ | UG 6-46 |
| 48. | $(\forall a)(\forall b)(\forall c)((a \subsetneq b \wedge b \subseteq c) \Rightarrow a \subsetneq c)$ | UG 5-47 |

□

**Theorem 2.36** $(\forall a)(\forall b)(\forall c)((a \subseteq c \wedge b \subseteq c) \Rightarrow a \cup b \subseteq c)$

***Proof***

| | | |
|---|---|---|
| 1. | $(\forall a)(\forall b)(a \subseteq b \Leftrightarrow (\forall c)(c \in a \Rightarrow c \in b))$ | Ax 2.9 |
| 2. | $(\forall a)(\forall b)(\forall c)(c \in a \cup b \Leftrightarrow (c \in a \vee c \in b))$ | T 2.18 |
| 3. | flag $d$ | FSUG |
| 4. | flag $e$ | FSUG |
| 5. | flag $f$ | FSUG |
| 6. | $d \subseteq f \wedge e \subseteq f$ | ACP |
| 7. | $d \subseteq f$ | S 6 |
| 8. | $(\forall b)(d \subseteq b \Leftrightarrow (\forall c)(c \in d \Rightarrow c \in b))$ | UI 1 |
| 9. | $d \subseteq f \Leftrightarrow (\forall c)(c \in d \Rightarrow c \in f)$ | UI 8 |
| 10. | $(\forall c)(c \in d \Rightarrow c \in f)$ | BEl 7, 9 |
| 11. | $e \subseteq f$ | S 6 |
| 12. | $(\forall b)(e \subseteq b \Leftrightarrow (\forall c)(c \in e \Rightarrow c \in b))$ | UI 1 |
| 13. | $e \subseteq f \Leftrightarrow (\forall c)(c \in e \Rightarrow c \in f)$ | UI 12 |
| 14. | $(\forall c)(c \in e \Rightarrow c \in f)$ | BEl 11, 13 |
| 15. | flag $g$ | FSUG |
| 16. | $g \in d \Rightarrow g \in f$ | UI 10 |
| 17. | $g \in e \Rightarrow g \in f$ | UI 14 |
| 18. | $g \in d \cup e$ | ACP |
| 19. | $(\forall b)(\forall c)(c \in d \cup b \Leftrightarrow (c \in d \vee c \in b))$ | UI 2 |
| 20. | $(\forall c)(c \in d \cup e \Leftrightarrow (c \in d \vee c \in e))$ | UI 19 |
| 21. | $g \in d \cup e \Leftrightarrow (g \in d \vee g \in e)$ | UI 20 |
| 22. | $g \in d \vee g \in e$ | BEl 18, 21 |
| 23. | $g \in f \vee g \in f$ | D 16, 17, 22 |
| 24. | $g \in f$ | Du 23 |
| 25. | $g \in d \cup e \Rightarrow g \in f$ | CP 18-24 |
| 26. | $(\forall c)(c \in d \cup e \Rightarrow c \in f)$ | UG 15-25 |
| 27. | $(\forall b)(d \cup e \subseteq b \Leftrightarrow (\forall c)(c \in d \cup e \Rightarrow c \in b))$ | UI 1 |
| 28. | $d \cup e \subseteq f \Leftrightarrow (\forall c)(c \in d \cup e \Rightarrow c \in f)$ | UI 27 |
| 29. | $d \cup e \subseteq f$ | BEl 26, 28 |
| 30. | $(d \subseteq f \wedge e \subseteq f) \Rightarrow d \cup e \subseteq f$ | CP 6-29 |
| 31. | $(\forall c)((d \subseteq c \wedge e \subseteq c) \Rightarrow d \cup e \subseteq c)$ | UG 5-30 |
| 32. | $(\forall b)(\forall c)((d \subseteq c \wedge b \subseteq c) \Rightarrow d \cup b \subseteq c)$ | UG 4-31 |
| 33. | $(\forall a)(\forall b)(\forall c)((a \subseteq c \wedge b \subseteq c) \Rightarrow a \cup b \subseteq c)$ | UG 3-32 |

□

**Theorem 2.37** $(\forall a)(\forall b)(a \in b \Rightarrow a \subseteq \cup b)$

***Proof***

| | | |
|---|---|---|
| 1. | $(\forall a)(\forall b)(b \in \cup a \Leftrightarrow (\exists c)(b \in c \wedge c \in a))$ | Ax 2.7 |
| 2. | $(\forall a)(\forall b)(a \subseteq b \Leftrightarrow (\forall c)(c \in a \Rightarrow c \in b))$ | Ax 2.9 |
| 3. | flag $d$ | FSUG |
| 4. | flag $e$ | FSUG |
| 5. | $d \in e$ | ACP |
| 6. | flag $f$ | FSUG |
| 7. | $f \in d$ | ACP |
| 8. | $f \in d \wedge d \in e$ | C 5, 7 |
| 9. | $(\exists c)(f \in c \wedge c \in e)$ | EG 8 |
| 10. | $(\forall b)(b \in \cup e \Leftrightarrow (\exists c)(b \in c \wedge c \in e))$ | UI 1 |
| 11. | $f \in \cup e \Leftrightarrow (\exists c)(f \in c \wedge c \in e)$ | UI 10 |
| 12. | $f \in \cup e$ | BEl 9, 11 |
| 13. | $f \in d \Rightarrow f \in \cup e$ | CP 7-12 |
| 14. | $(\forall c)(c \in d \Rightarrow c \in \cup e)$ | UG 6-13 |
| 15. | $(\forall b)(d \subseteq b \Leftrightarrow (\forall c)(c \in d \Rightarrow c \in b))$ | UI 2 |
| 16. | $d \subseteq \cup e \Leftrightarrow (\forall c)(c \in d \Rightarrow c \in \cup e)$ | UI 15 |
| 17. | $d \subseteq \cup e$ | BEl 14, 16 |
| 18. | $d \in e \Rightarrow d \subseteq \cup e$ | CP 5-17 |
| 19. | $(\forall b)(d \in b \Rightarrow d \subseteq \cup b)$ | UG 4-18 |
| 20. | $(\forall a)(\forall b)(a \in b \Rightarrow a \subseteq \cup b)$ | UG 3-19 |

□

**Theorem 2.38** $(\forall a)(\forall b)((\forall c)(c \in a \Rightarrow c \subseteq b) \Rightarrow \cup a \subseteq b)$

***Proof***

| | | |
|---|---|---|
| 1. | $(\forall a)(\forall b)(b \in \cup a \Leftrightarrow (\exists c)(b \in c \wedge c \in a))$ | Ax 2.7 |
| 2. | $(\forall a)(\forall b)(a \subseteq b \Leftrightarrow (\forall c)(c \in a \Rightarrow c \in b))$ | Ax 2.9 |
| 3. | flag $d$ | FSUG |
| 4. | flag $e$ | FSUG |
| 5. | $(\forall c)(c \in d \Rightarrow c \subseteq e)$ | ACP |
| 6. | flag $f$ | FSUG |
| 7. | $f \in \cup d$ | ACP |
| 8. | $(\forall b)(b \in \cup d \Leftrightarrow (\exists c)(b \in c \wedge c \in d))$ | UI 1 |
| 9. | $f \in \cup d \Leftrightarrow (\exists c)(f \in c \wedge c \in d)$ | UI 8 |
| 10. | $(\exists c)(f \in c \wedge c \in d)$ | BEl 7, 9 |
| 11. | $f \in g \wedge g \in d$ | EI 10, flag $g$ |
| 12. | $g \in d \Rightarrow g \subseteq e$ | UI 5 |
| 13. | $g \in d$ | S 11 |
| 14. | $g \subseteq e$ | MP 12, 13 |

| | | |
|---|---|---|
| 15. | $(\forall b)(g \subseteq b \Leftrightarrow (\forall c)(c \in g \Rightarrow c \in b))$ | UI 2 |
| 16. | $g \subseteq e \Leftrightarrow (\forall c)(c \in g \Rightarrow c \in e)$ | UI 15 |
| 17. | $(\forall c)(c \in g \Rightarrow c \in e)$ | BEI 14, 16 |
| 18. | $f \in g \Rightarrow f \in e$ | UI 17 |
| 19. | $f \in g$ | S 11 |
| 20. | $f \in e$ | MP 18, 19 |
| 21. | $f \in \cup d \Rightarrow f \in e$ | CP 7-20 |
| 22. | $(\forall c)(c \in \cup d \Rightarrow c \in e)$ | UG 6-21 |
| 23. | $(\forall b)(\cup d \subseteq b \Leftrightarrow (\forall c)(c \in \cup d \Rightarrow c \in b))$ | UI 2 |
| 24. | $\cup d \subseteq e \Leftrightarrow (\forall c)(c \in \cup d \Rightarrow c \in e)$ | UI 23 |
| 25. | $\cup d \subseteq e$ | BEI 22, 24 |
| 26. | $(\forall c)(c \in d \Rightarrow c \subseteq e) \Rightarrow \cup d \subseteq e$ | CP 5-25 |
| 27. | $(\forall b)((\forall c)(c \in d \Rightarrow c \subseteq b) \Rightarrow \cup d \subseteq b)$ | UG 4-26 |
| 28. | $(\forall a)(\forall b)((\forall c)(c \in a \Rightarrow c \subseteq b) \Rightarrow \cup a \subseteq b)$ | UG 3-27 |

□

**Theorem 2.39** $(\forall a)(\forall b)(a \subseteq b \Leftrightarrow a \cup b = b)$

***Proof***

| | | |
|---|---|---|
| 1. | $(\forall a)(\forall b)(b \subseteq a \cup b)$ | T 2.29 |
| 2. | $(\forall a)(a \subseteq a)$ | T 2.25 |
| 3. | $(\forall a)(\forall b)(\forall c)((a \subseteq c \wedge b \subseteq c) \Rightarrow a \cup b \subseteq c)$ | T 2.36 |
| 4. | $(\forall a)(\forall b)(a = b \Leftrightarrow (a \subseteq b \wedge b \subseteq a))$ | T 2.24 |
| 5. | $(\forall a)(\forall b)(a \subseteq a \cup b)$ | T 2.27 |
| 6. | flag $d$ | FSUG |
| 7. | flag $e$ | FSUG |
| 8. | $d \subseteq e$ | ACP |
| 9. | $(\forall b)(b \subseteq d \cup b)$ | UI 1 |
| 10. | $e \subseteq d \cup e$ | UI 9 |
| 11. | $e \subseteq e$ | UI 2 |
| 12. | $d \subseteq e \wedge e \subseteq e$ | C 8, 11 |
| 13. | $(\forall b)(\forall c)((d \subseteq c \wedge b \subseteq c) \Rightarrow d \cup b \subseteq c)$ | UI 3 |
| 14. | $(\forall c)((d \subseteq c \wedge e \subseteq c) \Rightarrow d \cup e \subseteq c)$ | UI 13 |
| 15. | $(d \subseteq e \wedge e \subseteq e) \Rightarrow d \cup e \subseteq e$ | UI 14 |
| 16. | $d \cup e \subseteq e$ | MP 12, 15 |
| 17. | $d \cup e \subseteq e \wedge e \subseteq d \cup e$ | C 10, 16 |
| 18. | $(\forall b)(d \cup e = b \Leftrightarrow (d \cup e \subseteq b \wedge b \subseteq d \cup e))$ | UI 4 |
| 19. | $d \cup e = e \Leftrightarrow (d \cup e \subseteq e \wedge e \subseteq d \cup e)$ | UI 18 |
| 20. | $d \cup e = e$ | BEI 17, 19 |
| 21. | $d \subseteq e \Rightarrow d \cup e = e$ | CP 8-20 |
| 22. | $d \cup e = e$ | ACP |

| | | |
|---|---|---|
| 23. | $(\forall b)(d \subseteq d \cup b)$ | UI 5 |
| 24. | $d \subseteq d \cup e$ | UI 23 |
| 25. | $d \subseteq e$ | IS 22, 24 |
| 26. | $d \cup e = e \Rightarrow d \subseteq e$ | CP 22-25 |
| 27. | $(d \subseteq e \Rightarrow d \cup e = e) \wedge (d \cup e = e \Rightarrow d \subseteq e)$ | C 21, 26 |
| 28. | $d \subseteq e \Leftrightarrow d \cup e = e$ | BE 27 |
| 29. | $(\forall b)(d \subseteq b \Rightarrow d \cup b = b)$ | UG 7-28 |
| 30. | $(\forall a)(\forall b)(a \subseteq b \Leftrightarrow a \cup b = b)$ | UG 6-29 |

□

## 2.4 Empty Set

**Axiom 2.12 (Definitions of Predicates Which If Applied to the Arguments $\alpha$ and $\beta$ Give the Atomic Formulas "$\alpha \notin \beta$", "$\alpha$ is not a member of $\beta$", "$\alpha$ is not an element of $\beta$" and "$\alpha$ does not belong to $\beta$")**

$(\forall a)(\forall b)(a \notin b \Leftrightarrow \neg(a \in b))$
$(\forall a)(\forall b)(a$ is not a member of $b \Leftrightarrow \neg(a \in b))$
$(\forall a)(\forall b)(a$ is not an element of $b \Leftrightarrow \neg(a \in b))$
$(\forall a)(\forall b)(a$ does not belong to $b \Leftrightarrow \neg(a \in b))$

**Theorem 2.40** $(\forall a)((\forall b)(b \notin a) \Rightarrow (\forall c)(a \subseteq c))$

***Proof***

| | | |
|---|---|---|
| 1. | $(\forall a)(\forall b)(a \subseteq b \Leftrightarrow (\forall c)(c \in a \Rightarrow c \in b))$ | Ax 2.9 |
| 2. | $(\forall a)(\forall b)(a \notin b \Leftrightarrow \neg(a \in b))$ | Ax 2.12 |
| 3. | flag $d$ | FSUG |
| 4. | $(\forall b)(b \notin d)$ | ACP |
| 5. | flag $e$ | FSUG |
| 6. | $(\forall b)(d \subseteq b \Leftrightarrow (\forall c)(c \in d \Rightarrow c \in b))$ | UI 1 |
| 7. | $d \subseteq e \Leftrightarrow (\forall c)(c \in d \Rightarrow c \in e)$ | UI 6 |
| 8. | $\neg(d \subseteq e)$ | AIP |
| 9. | $\neg(\forall c)(c \in d \Rightarrow c \in e)$ | NBE 7, 8 |
| 10. | $(\exists c)\neg(c \in d \Rightarrow c \in e)$ | QN 9 |
| 11. | $\neg(f \in d \Rightarrow f \in e)$ | EI 10, flag $f$ |
| 12. | $\neg((\neg(f \in d)) \vee f \in e)$ | CE 11 |
| 13. | $(\neg\neg(f \in d)) \wedge \neg(f \in e)$ | DM 12 |
| 14. | $f \in d \wedge \neg(f \in e)$ | DN 13 |
| 15. | $f \in d$ | S 14 |

| | | |
|---|---|---|
| 16. | $f \notin d$ | UI 4 |
| 17. | $(\forall b)(f \notin b \Leftrightarrow \neg(f \in b))$ | UI 2 |
| 18. | $f \notin d \Leftrightarrow \neg(f \in d)$ | UI 17 |
| 19. | $\neg(f \in d)$ | BEl 16, 18 |
| 20. | $f \in d \wedge \neg(f \in d)$ | C 15, 19 |
| 21. | $\neg\neg(d \subseteq e)$ | IP 8-20 |
| 22. | $d \subseteq e$ | DN 21 |
| 23. | $(\forall c)(d \subseteq c)$ | UG 5-22 |
| 24. | $(\forall b)(b \notin d) \Rightarrow (\forall c)(d \subseteq c)$ | CP 4-23 |
| 25. | $(\forall a)((\forall b)(b \notin a) \Rightarrow (\forall c)(a \subseteq c))$ | UG 3-24 |

□

**Axiom 2.13 (Zermelo Axiom: Axiom of the Empty Set)** $(\exists a)(\forall b)(b \notin a)$

**Theorem 2.41** $(\exists! a)(\forall b)(b \notin a)$

***Proof***

| | | |
|---|---|---|
| 1. | $(\exists a)(\forall b)(b \notin a)$ | Ax 2.13 |
| 2. | $(\forall a)((\forall b)(b \notin a) \Rightarrow (\forall c)(a \subseteq c))$ | T 2.40 |
| 3. | $(\forall a)(\forall b)(a = b \Leftrightarrow (a \subseteq b \wedge b \subseteq a))$ | T 2.24 |
| 4. | $(\forall b)(b \notin d)$ | EI 1, flag $d$ |
| 5. | $(\forall b)(b \notin d) \Rightarrow (\forall c)(d \subseteq c)$ | UI 2 |
| 6. | $(\forall c)(d \subseteq c)$ | MP 4, 5 |
| 7. | flag $e$ | FSUG |
| 8. | $(\forall b)(b \notin e)$ | ACP |
| 9. | $(\forall b)(b \notin e) \Rightarrow (\forall c)(e \subseteq c)$ | UI 2 |
| 10. | $(\forall c)(e \subseteq c)$ | MP 8, 9 |
| 11. | $e \subseteq d$ | UI 10 |
| 12. | $d \subseteq e$ | UI 6 |
| 13. | $e \subseteq d \wedge d \subseteq e$ | C 11, 12 |
| 14. | $(\forall b)(e = b \Leftrightarrow (e \subseteq b \wedge b \subseteq e))$ | UI 3 |
| 15. | $e = d \Leftrightarrow (e \subseteq d \wedge d \subseteq e)$ | UI 14 |
| 16. | $e = d$ | BEl 13, 15 |
| 17. | $(\forall b)(b \notin e) \Rightarrow e = d$ | CP 8-16 |
| 18. | $(\forall c)((\forall b)(b \notin c) \Rightarrow c = d)$ | UG 7-17 |
| 19. | $(\forall b)(b \notin d) \wedge (\forall c)((\forall b)(b \notin c) \Rightarrow c = d)$ | C 4, 18 |
| 20. | $(\exists a)((\forall b)(b \notin a) \wedge (\forall c)((\forall b)(b \notin c) \Rightarrow c = a))$ | EG 19 |
| 21. | $(\exists! a)(\forall b)(b \notin a)$ | UQ 20 |

□

**Axiom 2.14 (Definitions of Constants "$\phi$" and "empty set" from Theorem 2.41)**

$(\forall a)(a \notin \phi)$
$(\forall a)(a \notin \text{empty set})$

**Theorem 2.42** $(\forall a)(\phi \subseteq a)$

***Proof***

| | | |
|---|---|---|
| 1. | $(\forall c)((\forall b)(b \notin c) \Rightarrow (\forall a)(c \subseteq a))$ | T 2.40 |
| 2. | $(\forall b)(b \notin \phi)$ | Ax 2.14 |
| 3. | $(\forall b)(b \notin \phi) \Rightarrow (\forall a)(\phi \subseteq a)$ | UI 1 |
| 4. | $(\forall a)(\phi \subseteq a)$ | MP 2, 3 |

□

**Theorem 2.43** $(\forall a)(a \subseteq \phi \Rightarrow a = \phi)$

***Proof***

| | | |
|---|---|---|
| 1. | $(\forall a)(\phi \subseteq a)$ | T 2.42 |
| 2. | $(\forall a)(\forall b)(a = b \Leftrightarrow (a \subseteq b \wedge b \subseteq a))$ | T 2.24 |
| 3. | flag $c$ | FSUG |
| 4. | $c \subseteq \phi$ | ACP |
| 5. | $\phi \subseteq c$ | UI 1 |
| 6. | $c \subseteq \phi \wedge \phi \subseteq c$ | C 4, 5 |
| 7. | $(\forall b)(c = b \Leftrightarrow (c \subseteq b \wedge b \subseteq c))$ | UI 2 |
| 8. | $c = \phi \Leftrightarrow (c \subseteq \phi \wedge \phi \subseteq c)$ | UI 7 |
| 9. | $c = \phi$ | BEl 6, 8 |
| 10. | $c \subseteq \phi \Rightarrow c = \phi$ | CP 4-9 |
| 11. | $(\forall a)(a \subseteq \phi \Rightarrow a = \phi)$ | UG 3-10 |

□

**Theorem 2.44** $(\forall a)(a \cup \phi = a)$

***Proof***

| | | |
|---|---|---|
| 1. | $(\forall a)(\phi \subseteq a)$ | T 2.42 |
| 2. | $(\forall a)(\forall b)(a \subseteq b \Leftrightarrow a \cup b = b)$ | T 2.39 |
| 3. | $(\forall a)(\forall b)(a \cup b = b \cup a)$ | T 2.19 |
| 4. | flag $c$ | FSUG |
| 5. | $\phi \subseteq c$ | UI 1 |

| | | |
|---|---|---|
| 6. | $(\forall b)(\phi \subseteq b \Leftrightarrow \phi \cup b = b)$ | UI 2 |
| 7. | $\phi \subseteq c \Leftrightarrow \phi \cup c = c$ | UI 6 |
| 8. | $\phi \cup c = c$ | BEl 5, 7 |
| 9. | $(\forall b)(c \cup b = b \cup c)$ | UI 3 |
| 10. | $c \cup \phi = \phi \cup c$ | UI 9 |
| 11. | $c \cup \phi = c$ | IS 8, 10 |
| 12. | $(\forall a)(a \cup \phi = a)$ | UG 4-11 |

□

**Theorem 2.45** $(\forall a)((\forall b)(b \notin a) \Rightarrow a = \phi)$

***Proof***

| | | |
|---|---|---|
| 1. | $(\exists! a)(\forall b)(b \notin a)$ | T 2.41 |
| 2. | $(\forall b)(b \notin \phi)$ | Ax 2.14 |
| 3. | $(\exists a)(\forall b)(b \notin a) \wedge$ $(\forall c)(\forall d)(((\forall b)(b \notin c) \wedge (\forall b)(b \notin d)) \Rightarrow c = d)$ | UQ 1 |
| 4. | $(\forall c)(\forall d)(((\forall b)(b \notin c) \wedge (\forall b)(b \notin d)) \Rightarrow c = d)$ | S 3 |
| 5. | $(\forall d)(((\forall b)(b \notin \phi) \wedge (\forall b)(b \notin d)) \Rightarrow \phi = d)$ | UI 4 |
| 6. | flag $e$ | FSUG |
| 7. | $((\forall b)(b \notin \phi) \wedge (\forall b)(b \notin e)) \Rightarrow \phi = e$ | UI 5 |
| 8. | $(\forall b)(b \notin e)$ | ACP |
| 9. | $(\forall b)(b \notin \phi) \wedge (\forall b)(b \notin e)$ | C 2, 8 |
| 10. | $\phi = e$ | MP 7, 9 |
| 11. | $e = \phi$ | ISy 10 |
| 12. | $(\forall b)(b \notin e) \Rightarrow e = \phi$ | CP 8-11 |
| 13. | $(\forall a)((\forall b)(b \notin a) \Rightarrow a = \phi)$ | UG 6-12 |

□

**Theorem 2.46** $(\forall a)(a \neq \phi \Rightarrow (\exists b)(b \in a))$

***Proof***

| | | |
|---|---|---|
| 1. | $(\forall a)((\forall b)(b \notin a) \Rightarrow a = \phi)$ | T 2.45 |
| 2. | $(\forall a)(\forall b)(a \notin b \Leftrightarrow \neg(a \in b))$ | Ax 2.12 |
| 3. | $(\forall a)(\forall b)(a \neq b \Leftrightarrow \neg(a = b))$ | Ax 2.5 |
| 4. | flag $c$ | FSUG |
| 5. | $(\forall b)(b \notin c) \Rightarrow c = \phi$ | UI 1 |
| 6. | $(\forall b)\neg(b \in c)$ | ACP |
| 7. | flag $d$ | FSUG |
| 8. | $\neg(d \in c)$ | UI 6 |

| | | |
|---|---|---|
| 9. | $(\forall b)(d \notin b \Leftrightarrow \neg(d \in b))$ | UI 2 |
| 10. | $d \notin c \Leftrightarrow \neg(d \in c)$ | UI 9 |
| 11. | $d \notin c$ | BEl 8, 10 |
| 12. | $(\forall b)(b \notin c)$ | UG 7-11 |
| 13. | $(\forall b)\neg(b \in c) \Rightarrow (\forall b)(b \notin c)$ | CP 6-12 |
| 14. | $(\forall b)\neg(b \in c) \Rightarrow c = \phi$ | HS 5, 13 |
| 15. | $(\neg(c = \phi)) \Rightarrow \neg(\forall b)\neg(b \in c)$ | Con 14 |
| 16. | $\neg(c = \phi)$ | ACP |
| 17. | $\neg(\forall b)\neg(b \in c)$ | MP 15, 16 |
| 18. | $(\exists b)(b \in c)$ | QN 17 |
| 19. | $(\neg(c = \phi)) \Rightarrow (\exists b)(b \in c)$ | CP 16-18 |
| 20. | $c \neq \phi$ | ACP |
| 21. | $(\forall b)(c \neq b \Leftrightarrow \neg(c = b))$ | UI 3 |
| 22. | $c \neq \phi \Leftrightarrow \neg(c = \phi)$ | UI 21 |
| 23. | $\neg(c = \phi)$ | BEl 20, 22 |
| 24. | $c \neq \phi \Rightarrow \neg(c = \phi)$ | CP 20-23 |
| 25. | $c \neq \phi \Rightarrow (\exists b)(b \in c)$ | HS 19, 24 |
| 26. | $(\forall a)(a \neq \phi \Rightarrow (\exists b)(b \in a))$ | UG 4-25 |

□

**Theorem 2.47** $(\forall a)(a = \phi \Rightarrow (\forall b)(b \notin a))$

***Proof***

| | | |
|---|---|---|
| 1. | $(\forall a)(a \notin \phi)$ | Ax 2.14 |
| 2. | flag $c$ | FSUG |
| 3. | $c = \phi$ | ACP |
| 4. | flag $d$ | FSUG |
| 5. | $d \notin \phi$ | UI 1 |
| 6. | $d \notin c$ | IS 3, 5 |
| 7. | $(\forall b)(b \notin c)$ | UG 4-6 |
| 8. | $c = \phi \Rightarrow (\forall b)(b \notin c)$ | CP 3-7 |
| 9. | $(\forall a)(a = \phi \Rightarrow (\forall b)(b \notin a))$ | UG 2-8 |

□

**Theorem 2.48** $(\forall a)((\exists b)(b \in a) \Rightarrow a \neq \phi)$

***Proof***

| | | |
|---|---|---|
| 1. | $(\forall a)(a = \phi \Rightarrow (\forall b)(b \notin a))$ | T 2.47 |
| 2. | $(\forall a)(\forall b)(a \notin b \Leftrightarrow \neg(a \in b))$ | Ax 2.12 |
| 3. | $(\forall a)(\forall b)(a \neq b \Leftrightarrow \neg(a = b))$ | Ax 2.5 |
| 4. | flag $c$ | FSUG |
| 5. | $(\exists b)(b \in c)$ | ACP |
| 6. | $d \in c$ | EI 5, flag $d$ |
| 7. | $c = \phi$ | AIP |
| 8. | $c = \phi \Rightarrow (\forall b)(b \notin c)$ | UI 1 |
| 9. | $(\forall b)(b \notin c)$ | MP 7, 8 |
| 10. | $d \notin c$ | UI 9 |
| 11. | $(\forall b)(d \notin b \Leftrightarrow \neg(d \in b))$ | UI 2 |
| 12. | $d \notin c \Leftrightarrow \neg(d \in c)$ | UI 11 |
| 13. | $\neg(d \in c)$ | BEl 10, 12 |
| 14. | $d \in c \wedge \neg(d \in c)$ | C 6, 13 |
| 15. | $\neg(c = \phi)$ | IP 7-14 |
| 16. | $(\forall b)(c \neq b \Leftrightarrow \neg(c = b))$ | UI 3 |
| 17. | $c \neq \phi \Leftrightarrow \neg(c = \phi)$ | UI 16 |
| 18. | $c \neq \phi$ | BEl 15, 17 |
| 19. | $(\exists b)(b \in c) \Rightarrow c \neq \phi$ | CP 5-18 |
| 20. | $(\forall a)((\exists b)(b \in a) \Rightarrow a \neq \phi)$ | UG 4-19 |

□

**Theorem 2.49** $\cup \phi = \phi$

***Proof***

| | | |
|---|---|---|
| 1. | $(\forall a)(\forall b)(b \in \cup a \Leftrightarrow (\exists c)(b \in c \wedge c \in a))$ | Ax 2.7 |
| 2. | $(\forall a)(a \notin \phi)$ | Ax 2.14 |
| 3. | $(\forall a)(\forall b)(a \notin b \Leftrightarrow \neg(a \in b))$ | Ax 2.12 |
| 4. | $(\forall a)((\forall b)(b \notin a) \Rightarrow a = \phi)$ | T 2.45 |
| 5. | flag $d$ | FSUG |
| 6. | $d \in \cup\phi$ | AIP |
| 7. | $(\forall b)(b \in \cup\phi \Leftrightarrow (\exists c)(b \in c \wedge c \in \phi))$ | UI 1 |
| 8. | $d \in \cup\phi \Leftrightarrow (\exists c)(d \in c \wedge c \in \phi)$ | UI 7 |
| 9. | $(\exists c)(d \in c \wedge c \in \phi)$ | BEl 6, 8 |
| 10. | $d \in e \wedge e \in \phi$ | EI 9, flag $e$ |
| 11. | $e \in \phi$ | S 10 |
| 12. | $e \notin \phi$ | UI 2 |
| 13. | $(\forall b)(e \notin b \Leftrightarrow \neg(e \in b))$ | UI 3 |
| 14. | $e \notin \phi \Leftrightarrow \neg(e \in \phi)$ | UI 13 |

| | | |
|---|---|---|
| 15. | $\neg(e \in \phi)$ | BEl 12, 14 |
| 16. | $e \in \phi \wedge \neg(e \in \phi)$ | C 11, 15 |
| 17. | $\neg(d \in \cup\phi)$ | IP 6-16 |
| 18. | $(\forall b)(d \notin b \Leftrightarrow \neg(d \in b))$ | UI 3 |
| 19. | $d \notin \cup\phi \Leftrightarrow \neg(d \in \cup\phi)$ | UI 18 |
| 20. | $d \notin \cup\phi$ | BEl 17, 19 |
| 21. | $(\forall b)(b \notin \cup\phi)$ | UG 5-20 |
| 22. | $(\forall b)(b \notin \cup\phi) \Rightarrow \cup\phi = \phi$ | UI 4 |
| 23. | $\cup\,\phi = \phi$ | MP 21, 22 |

□

## 2.5 Intersection

**Axiom Schema 2.1 (Zermelo Axiom Schema: Axiom Schema of Separation)** $(\forall a)(\forall a_1)(\forall a_2)...(\forall a_n)(\exists b)(\forall c)(c \in b \Leftrightarrow (c \in a \wedge A(c, a_1, a_2, ..., a_n)))$

**Theorem 2.50** $(\forall a)(\forall b)(\exists! c)(\forall d)(d \in c \Leftrightarrow (d \in a \wedge d \in b))$

***Proof***

| | | |
|---|---|---|
| 1. | $(\forall a)(\forall b)(\exists c)(\forall d)(d \in c \Leftrightarrow (d \in a \wedge d \in b))$ | AS 2.1 |
| 2. | $(\forall a)(\forall b)((\forall d)(d \in a \Leftrightarrow d \in b) \Rightarrow a = b)$ | Ax 2.1 |
| 3. | flag $e$ | FSUG |
| 4. | $(\forall b)(\exists c)(\forall d)(d \in c \Leftrightarrow (d \in e \wedge d \in b))$ | UI 1 |
| 5. | flag $f$ | FSUG |
| 6. | $(\exists c)(\forall d)(d \in c \Leftrightarrow (d \in e \wedge d \in f))$ | UI 4 |
| 7. | flag $g$ | FSUG |
| 8. | flag $h$ | FSUG |
| 9. | $(\forall d)(d \in g \Leftrightarrow (d \in e \wedge d \in f)) \wedge$ $(\forall d)(d \in h \Leftrightarrow (d \in e \wedge d \in f))$ | ACP |
| 10. | $(\forall d)(d \in g \Leftrightarrow (d \in e \wedge d \in f))$ | S 9 |
| 11. | $(\forall d)(d \in h \Leftrightarrow (d \in e \wedge d \in f))$ | S 9 |
| 12. | flag $i$ | FSUG |
| 13. | $i \in g \Leftrightarrow (i \in e \wedge i \in f)$ | UI 10 |
| 14. | $i \in h \Leftrightarrow (i \in e \wedge i \in f)$ | UI 11 |
| 15. | $i \in g \Leftrightarrow i \in h$ | BT 13, 14 |
| 16. | $(\forall d)(d \in g \Leftrightarrow d \in h)$ | UG 12-15 |
| 17. | $(\forall b)((\forall d)(d \in g \Leftrightarrow d \in b) \Rightarrow g = b)$ | UI 2 |
| 18. | $(\forall d)(d \in g \Leftrightarrow d \in h) \Rightarrow g = h$ | UI 17 |
| 19. | $g = h$ | MP 16, 18 |
| 20. | $((\forall d)(d \in g \Leftrightarrow (d \in e \wedge d \in f)) \wedge$ $(\forall d)(d \in h \Leftrightarrow (d \in e \wedge d \in f))) \Rightarrow g = h$ | CP 9-19 |

21. $(\forall j)(((\forall d)(d \in g \Leftrightarrow (d \in e \wedge d \in f))\wedge$
$(\forall d)(d \in j \Leftrightarrow (d \in e \wedge d \in f))) \Rightarrow g = j)$ UG 8-20
22. $(\forall k)(\forall j)(((\forall d)(d \in k \Leftrightarrow (d \in e \wedge d \in f))\wedge$
$(\forall d)(d \in j \Leftrightarrow (d \in e \wedge d \in f))) \Rightarrow k = j)$ UG 7-21
23. $(\exists c)(\forall d)(d \in c \Leftrightarrow (d \in e \wedge d \in f))\wedge$
$(\forall k)(\forall j)(((\forall d)(d \in k \Leftrightarrow (d \in e \wedge d \in f))\wedge$
$(\forall d)(d \in j \Leftrightarrow (d \in e \wedge d \in f))) \Rightarrow k = j)$ C 6, 22
24. $(\exists! c)(\forall d)(d \in c \Leftrightarrow (d \in e \wedge d \in f))$ UQ 23
25. $(\forall b)(\exists! c)(\forall d)(d \in c \Leftrightarrow (d \in e \wedge d \in b))$ UG 5-24
26. $(\forall a)(\forall b)(\exists! c)(\forall d)(d \in c \Leftrightarrow (d \in a \wedge d \in b))$ UG 3-25

□

**Axiom 2.15 (Definitions of Operations, from Theorem 2.50, Which If Applied to the Arguments $\alpha$ and $\beta$ Give the Terms "$\alpha \cap \beta$" and "intersection of $\alpha$ and $\beta$")**

$(\forall a)(\forall b)(\forall c)(c \in a \cap b \Leftrightarrow (c \in a \wedge c \in b))$
$(\forall a)(\forall b)(\forall c)(c \in \text{intersection of } a \text{ and } b \Leftrightarrow (c \in a \wedge c \in b))$

**Theorem 2.51** $(\forall a)(\forall b)(a \cap b = b \cap a)$

***Proof***

| | | |
|---|---|---|
| 1. | $(\forall a)(\forall b)(\forall c)(c \in a \cap b \Leftrightarrow (c \in a \wedge c \in b))$ | Ax 2.15 |
| 2. | $(\forall a)(\forall b)((\forall c)(c \in a \Leftrightarrow c \in b) \Rightarrow a = b)$ | Ax 2.1 |
| 3. | flag $d$ | FSUG |
| 4. | $(\forall b)(\forall c)(c \in d \cap b \Leftrightarrow (c \in d \wedge c \in b))$ | UI 1 |
| 5. | flag $e$ | FSUG |
| 6. | $(\forall c)(c \in d \cap e \Leftrightarrow (c \in d \wedge c \in e))$ | UI 4 |
| 7. | flag $f$ | FSUG |
| 8. | $f \in d \cap e \Leftrightarrow (f \in d \wedge f \in e)$ | UI 6 |
| 9. | $f \in d \cap e \Leftrightarrow (f \in e \wedge f \in d)$ | Co 8 |
| 10. | $(\forall b)(\forall c)(c \in e \cap b \Leftrightarrow (c \in e \wedge c \in b))$ | UI 1 |
| 11. | $(\forall c)(c \in e \cap d \Leftrightarrow (c \in e \wedge c \in d))$ | UI 10 |
| 12. | $f \in e \cap d \Leftrightarrow (f \in e \wedge f \in d)$ | UI 11 |
| 13. | $f \in d \cap e \Leftrightarrow f \in e \cap d$ | BT 9, 12 |
| 14. | $(\forall c)(c \in d \cap e \Leftrightarrow c \in e \cap d)$ | UG 7-13 |
| 15. | $(\forall b)((\forall c)(c \in d \cap e \Leftrightarrow c \in b) \Rightarrow d \cap e = b)$ | UI 2 |
| 16. | $(\forall c)(c \in d \cap e \Leftrightarrow c \in e \cap d) \Rightarrow d \cap e = e \cap d$ | UI 15 |
| 17. | $d \cap e = e \cap d$ | MP 14, 16 |
| 18. | $(\forall b)(d \cap b = b \cap d)$ | UG 5-17 |
| 19. | $(\forall a)(\forall b)(a \cap b = b \cap a)$ | UG 3-18 |

□

**Theorem 2.52** $(\forall a)(\forall b)(\forall c)(\forall d)(d \in a \cap (b \cap c) \Leftrightarrow (d \in a \wedge (d \in b \wedge d \in c)))$

***Proof***

| | | |
|---|---|---|
| 1. | $(\forall a)(\forall b)(\forall c)(c \in a \cap b \Leftrightarrow (c \in a \wedge c \in b))$ | Ax 2.15 |
| 2. | flag $e$ | FSUG |
| 3. | flag $f$ | FSUG |
| 4. | flag $g$ | FSUG |
| 5. | flag $h$ | FSUG |
| 6. | $(\forall b)(\forall c)(c \in e \cap b \Leftrightarrow (c \in e \wedge c \in b))$ | UI 1 |
| 7. | $(\forall c)(c \in e \cap (f \cap g) \Leftrightarrow (c \in e \wedge c \in f \cap g))$ | UI 6 |
| 8. | $h \in e \cap (f \cap g) \Leftrightarrow (h \in e \wedge h \in f \cap g)$ | UI 7 |
| 9. | $(h \in e \cap (f \cap g) \Rightarrow (h \in e \wedge h \in f \cap g)) \wedge$ $((h \in e \wedge h \in f \cap g) \Rightarrow h \in e \cap (f \cap g))$ | BE 8 |
| 10. | $h \in e \cap (f \cap g) \Rightarrow (h \in e \wedge h \in f \cap g)$ | S 9 |
| 11. | $h \in e \cap (f \cap g)$ | ACP |
| 12. | $h \in e \wedge h \in f \cap g$ | MP 10, 11 |
| 13. | $h \in e$ | S 12 |
| 14. | $h \in f \cap g$ | S 12 |
| 15. | $(\forall b)(\forall c)(c \in f \cap b \Leftrightarrow (c \in f \wedge c \in b))$ | UI 1 |
| 16. | $(\forall c)(c \in f \cap g \Leftrightarrow (c \in f \wedge c \in g))$ | UI 15 |
| 17. | $h \in f \cap g \Leftrightarrow (h \in f \wedge h \in g)$ | UI 16 |
| 18. | $h \in f \wedge h \in g$ | BEl 14, 17 |
| 19. | $h \in e \wedge (h \in f \wedge h \in g)$ | C 13, 18 |
| 20. | $h \in e \cap (f \cap g) \Rightarrow (h \in e \wedge (h \in f \wedge h \in g))$ | CP 11-19 |
| 21. | $h \in e \wedge (h \in f \wedge h \in g)$ | ACP |
| 22. | $h \in e$ | S 21 |
| 23. | $h \in f \wedge h \in g$ | S 21 |
| 24. | $(\forall b)(\forall c)(c \in f \cap b \Leftrightarrow (c \in f \wedge c \in b))$ | UI 1 |
| 25. | $(\forall c)(c \in f \cap g \Leftrightarrow (c \in f \wedge c \in g))$ | UI 24 |
| 26. | $h \in f \cap g \Leftrightarrow (h \in f \wedge h \in g)$ | UI 25 |
| 27. | $h \in f \cap g$ | BEl 23, 26 |
| 28. | $h \in e \wedge h \in f \cap g$ | C 22, 27 |
| 29. | $h \in e \cap (f \cap g)$ | BEl 8, 28 |
| 30. | $(h \in e \wedge (h \in f \wedge h \in g)) \Rightarrow h \in e \cap (f \cap g)$ | CP 21-29 |
| 31. | $(h \in e \cap (f \cap g) \Rightarrow (h \in e \wedge (h \in f \wedge h \in g))) \wedge$ $((h \in e \wedge (h \in f \wedge h \in g)) \Rightarrow h \in e \cap (f \cap g))$ | C 20, 30 |
| 32. | $h \in e \cap (f \cap g) \Leftrightarrow (h \in e \wedge (h \in f \wedge h \in g))$ | BE 31 |
| 33. | $(\forall d)(d \in e \cap (f \cap g) \Leftrightarrow (d \in e \wedge (d \in f \wedge d \in g)))$ | UG 5-32 |
| 34. | $(\forall c)(\forall d)(d \in e \cap (f \cap c) \Leftrightarrow (d \in e \wedge (d \in f \wedge d \in c)))$ | UG 4-33 |
| 35. | $(\forall b)(\forall c)(\forall d)(d \in e \cap (b \cap c) \Leftrightarrow (d \in e \wedge (d \in b \wedge d \in c)))$ | UG 3-34 |
| 36. | $(\forall a)(\forall b)(\forall c)(\forall d)(d \in a \cap (b \cap c) \Leftrightarrow (d \in a \wedge (d \in b \wedge d \in c)))$ | UG 2-35 |

□

**Theorem 2.53** $(\forall a)(\forall b)(\forall c)(a \cap (b \cap c) = (a \cap b) \cap c)$

***Proof***

| | | |
|---|---|---|
| 1. | $(\forall a)(\forall b)(\forall c)(\forall d)(d \in a \cap (b \cap c) \Leftrightarrow (d \in a \wedge (d \in b \wedge d \in c)))$ | T 2.52 |
| 2. | $(\forall a)(\forall b)(a \cap b = b \cap a)$ | T 2.51 |
| 3. | $(\forall a)(\forall b)((\forall c)(c \in a \Leftrightarrow c \in b) \Rightarrow a = b)$ | Ax 2.1 |
| 4. | flag $e$ | FSUG |
| 5. | flag $f$ | FSUG |
| 6. | flag $g$ | FSUG |
| 7. | flag $h$ | FSUG |
| 8. | $(\forall b)(\forall c)(\forall d)(d \in e \cap (b \cap c) \Leftrightarrow$ $(d \in e \wedge (d \in b \wedge d \in c)))$ | UI 1 |
| 9. | $(\forall c)(\forall d)(d \in e \cap (f \cap c) \Leftrightarrow$ $(d \in e \wedge (d \in f \wedge d \in c)))$ | UI 8 |
| 10. | $(\forall d)(d \in e \cap (f \cap g) \Leftrightarrow (d \in e \wedge (d \in f \wedge d \in g)))$ | UI 9 |
| 11. | $h \in e \cap (f \cap g) \Leftrightarrow (h \in e \wedge (h \in f \wedge h \in g))$ | UI 10 |
| 12. | $h \in e \cap (f \cap g) \Leftrightarrow ((h \in e \wedge h \in f) \wedge h \in g)$ | A 11 |
| 13. | $h \in e \cap (f \cap g) \Leftrightarrow (h \in g \wedge (h \in e \wedge h \in f))$ | Co 12 |
| 14. | $(\forall b)(\forall c)(\forall d)(d \in g \cap (b \cap c) \Leftrightarrow$ $(d \in g \wedge (d \in b \wedge d \in c)))$ | UI 1 |
| 15. | $(\forall c)(\forall d)(d \in g \cap (e \cap c) \Leftrightarrow$ $(d \in g \wedge (d \in e \wedge d \in c)))$ | UI 14 |
| 16. | $(\forall d)(d \in g \cap (e \cap f) \Leftrightarrow (d \in g \wedge (d \in e \wedge d \in f)))$ | UI 15 |
| 17. | $h \in g \cap (e \cap f) \Leftrightarrow (h \in g \wedge (h \in e \wedge h \in f))$ | UI 16 |
| 18. | $h \in e \cap (f \cap g) \Leftrightarrow h \in g \cap (e \cap f)$ | BT 13, 17 |
| 19. | $(\forall b)(g \cap b = b \cap g)$ | UI 2 |
| 20. | $g \cap (e \cap f) = (e \cap f) \cap g$ | UI 19 |
| 21. | $h \in e \cap (f \cap g) \Leftrightarrow h \in (e \cap f) \cap g$ | IS 18, 20 |
| 22. | $(\forall c)(c \in e \cap (f \cap g) \Leftrightarrow c \in (e \cap f) \cap g)$ | UG 7-21 |
| 23. | $(\forall b)((\forall c)(c \in e \cap (f \cap g) \Leftrightarrow c \in b) \Rightarrow$ $e \cap (f \cap g) = b)$ | UI 3 |
| 24. | $(\forall c)(c \in e \cap (f \cap g) \Leftrightarrow c \in (e \cap f) \cap g)$ $\Rightarrow e \cap (f \cap g) = (e \cap f) \cap g$ | UI 23 |
| 25. | $e \cap (f \cap g) = (e \cap f) \cap g$ | MP 22, 24 |
| 26. | $(\forall c)(e \cap (f \cap c) = (e \cap f) \cap c)$ | UG 6-25 |
| 27. | $(\forall b)(\forall c)(e \cap (b \cap c) = (e \cap b) \cap c)$ | UG 5-26 |
| 28. | $(\forall a)(\forall b)(\forall c)(a \cap (b \cap c) = (a \cap b) \cap c)$ | UG 4-27 |

□

**Theorem 2.54** $(\forall a)(a \cap a = a)$

***Proof***

| | | |
|---|---|---|
| 1. | $(\forall a)(\forall b)(\forall c)(c \in a \cap b \Leftrightarrow (c \in a \wedge c \in b))$ | Ax 2.15 |
| 2. | $(\forall a)(\forall b)((\forall c)(c \in a \Leftrightarrow c \in b) \Rightarrow a = b)$ | Ax 2.1 |
| 3. | flag $d$ | FSUG |
| 4. | flag $e$ | FSUG |
| 5. | $(\forall b)(\forall c)(c \in d \cap b \Leftrightarrow (c \in d \wedge c \in b))$ | UI 1 |
| 6. | $(\forall c)(c \in d \cap d \Leftrightarrow (c \in d \wedge c \in d))$ | UI 5 |
| 7. | $e \in d \cap d \Leftrightarrow (e \in d \wedge e \in d)$ | UI 6 |
| 8. | $e \in d \cap d \Leftrightarrow e \in d$ | Du 7 |
| 9. | $(\forall c)(c \in d \cap d \Leftrightarrow c \in d)$ | UG 4-8 |
| 10. | $(\forall b)((\forall c)(c \in d \cap d \Leftrightarrow c \in b) \Rightarrow d \cap d = b)$ | UI 2 |
| 11. | $(\forall c)(c \in d \cap d \Leftrightarrow c \in d) \Rightarrow d \cap d = d$ | UI 10 |
| 12. | $d \cap d = d$ | MP 9, 11 |
| 13. | $(\forall a)(a \cap a = a)$ | UG 3-12 |

□

**Theorem 2.55** $(\forall a)(\forall b)(a \cap b \subseteq a)$

***Proof***

| | | |
|---|---|---|
| 1. | $(\forall a)(\forall b)(\forall c)(c \in a \cap b \Leftrightarrow (c \in a \wedge c \in b))$ | Ax 2.15 |
| 2. | $(\forall a)(\forall b)(a \subseteq b \Leftrightarrow (\forall c)(c \in a \Rightarrow c \in b))$ | Ax 2.9 |
| 3. | flag $d$ | FSUG |
| 4. | flag $e$ | FSUG |
| 5. | flag $f$ | FSUG |
| 6. | $f \in d \cap e$ | ACP |
| 7. | $(\forall b)(\forall c)(c \in d \cap b \Leftrightarrow (c \in d \wedge c \in b))$ | UI 1 |
| 8. | $(\forall c)(c \in d \cap e \Leftrightarrow (c \in d \wedge c \in e))$ | UI 7 |
| 9. | $f \in d \cap e \Leftrightarrow (f \in d \wedge f \in e)$ | UI 8 |
| 10. | $f \in d \wedge f \in e$ | BEl 6, 9 |
| 11. | $f \in d$ | S 10 |
| 12. | $f \in d \cap e \Rightarrow f \in d$ | CP 6-11 |
| 13. | $(\forall c)(c \in d \cap e \Rightarrow c \in d)$ | UG 5-12 |
| 14. | $(\forall b)(d \cap e \subseteq b \Leftrightarrow (\forall c)(c \in d \cap e \Rightarrow c \in b))$ | UI 2 |
| 15. | $d \cap e \subseteq d \Leftrightarrow (\forall c)(c \in d \cap e \Rightarrow c \in d)$ | UI 14 |
| 16. | $d \cap e \subseteq d$ | BEl 13, 15 |
| 17. | $(\forall b)(d \cap b \subseteq d)$ | UG 4-16 |
| 18. | $(\forall a)(\forall b)(a \cap b \subseteq a)$ | UG 3-17 |

□

**Theorem 2.56** $(\forall a)(\forall b)(a \cap b \subseteq b)$

***Proof***

| | | |
|---|---|---|
| 1. | $(\forall a)(\forall b)(\forall c)(c \in a \cap b \Leftrightarrow (c \in a \wedge c \in b))$ | Ax 2.15 |
| 2. | $(\forall a)(\forall b)(a \subseteq b \Leftrightarrow (\forall c)(c \in a \Rightarrow c \in b))$ | Ax 2.9 |
| 3. | $\quad$ flag $d$ | FSUG |
| 4. | $\quad\quad$ flag $e$ | FSUG |
| 5. | $\quad\quad\quad$ flag $f$ | FSUG |
| 6. | $\quad\quad\quad\quad f \in d \cap e$ | ACP |
| 7. | $\quad\quad\quad\quad (\forall b)(\forall c)(c \in d \cap b \Leftrightarrow (c \in d \wedge c \in b))$ | UI 1 |
| 8. | $\quad\quad\quad\quad (\forall c)(c \in d \cap e \Leftrightarrow (c \in d \wedge c \in e))$ | UI 7 |
| 9. | $\quad\quad\quad\quad f \in d \cap e \Leftrightarrow (f \in d \wedge f \in e)$ | UI 8 |
| 10. | $\quad\quad\quad\quad f \in d \wedge f \in e$ | BEl 6, 9 |
| 11. | $\quad\quad\quad\quad f \in e$ | S 10 |
| 12. | $\quad\quad\quad f \in d \cap e \Rightarrow f \in e$ | CP 6-11 |
| 13. | $\quad\quad (\forall c)(c \in d \cap e \Rightarrow c \in e)$ | UG 5-12 |
| 14. | $\quad\quad (\forall b)(d \cap e \subseteq b \Leftrightarrow (\forall c)(c \in d \cap e \Rightarrow c \in b))$ | UI 2 |
| 15. | $\quad\quad d \cap e \subseteq e \Leftrightarrow (\forall c)(c \in d \cap e \Rightarrow c \in e)$ | UI 14 |
| 16. | $\quad\quad d \cap e \subseteq e$ | BEl 13, 15 |
| 17. | $\quad (\forall b)(d \cap b \subseteq b)$ | UG 4-16 |
| 18. | $(\forall a)(\forall b)(a \cap b \subseteq b)$ | UG 3-17 |

□

**Theorem 2.57** $(\forall a)(\forall b)(\forall c)((c \subseteq a \wedge c \subseteq b) \Rightarrow c \subseteq a \cap b)$

***Proof***

| | | |
|---|---|---|
| 1. | $(\forall a)(\forall b)(a \subseteq b \Leftrightarrow (\forall c)(c \in a \Rightarrow c \in b))$ | Ax 2.9 |
| 2. | $(\forall a)(\forall b)(\forall c)(c \in a \cap b \Leftrightarrow (c \in a \wedge c \in b))$ | Ax 2.15 |
| 3. | $\quad$ flag $d$ | FSUG |
| 4. | $\quad\quad$ flag $e$ | FSUG |
| 5. | $\quad\quad\quad$ flag $f$ | FSUG |
| 6. | $\quad\quad\quad\quad f \subseteq d \wedge f \subseteq e$ | ACP |
| 7. | $\quad\quad\quad\quad f \subseteq d$ | S 6 |
| 8. | $\quad\quad\quad\quad (\forall b)(f \subseteq b \Leftrightarrow (\forall c)(c \in f \Rightarrow c \in b))$ | UI 1 |
| 9. | $\quad\quad\quad\quad f \subseteq d \Leftrightarrow (\forall c)(c \in f \Rightarrow c \in d)$ | UI 8 |
| 10. | $\quad\quad\quad\quad (\forall c)(c \in f \Rightarrow c \in d)$ | BEl 7, 9 |
| 11. | $\quad\quad\quad\quad f \subseteq e$ | S 6 |
| 12. | $\quad\quad\quad\quad f \subseteq e \Leftrightarrow (\forall c)(c \in f \Rightarrow c \in e)$ | UI 8 |
| 13. | $\quad\quad\quad\quad (\forall c)(c \in f \Rightarrow c \in e)$ | BEl 11, 12 |
| 14. | $\quad\quad\quad\quad\quad$ flag $g$ | FSUG |

| | | |
|---|---|---|
| 15. | $g \in f \Rightarrow g \in d$ | UI 10 |
| 16. | $g \in f \Rightarrow g \in e$ | UI 13 |
| 17. | $g \in f$ | ACP |
| 18. | $g \in d$ | MP 15, 17 |
| 19. | $g \in e$ | MP 16, 17 |
| 20. | $g \in d \wedge g \in e$ | C 18, 19 |
| 21. | $(\forall b)(\forall c)(c \in d \cap b \Leftrightarrow (c \in d \wedge c \in b))$ | UI 2 |
| 22. | $(\forall c)(c \in d \cap e \Leftrightarrow (c \in d \wedge c \in e))$ | UI 21 |
| 23. | $g \in d \cap e \Leftrightarrow (g \in d \wedge g \in e)$ | UI 22 |
| 24. | $g \in d \cap e$ | BEl 20, 23 |
| 25. | $g \in f \Rightarrow g \in d \cap e$ | CP 17-24 |
| 26. | $(\forall c)(c \in f \Rightarrow c \in d \cap e)$ | UG 14-25 |
| 27. | $f \subseteq d \cap e \Leftrightarrow (\forall c)(c \in f \Rightarrow c \in d \cap e)$ | UI 8 |
| 28. | $f \subseteq d \cap e$ | BEl 26, 27 |
| 29. | $(f \subseteq d \wedge f \subseteq e) \Rightarrow f \subseteq d \cap e$ | CP 6-28 |
| 30. | $(\forall c)((c \subseteq d \wedge c \subseteq e) \Rightarrow c \subseteq d \cap e)$ | UG 5-29 |
| 31. | $(\forall b)(\forall c)((c \subseteq d \wedge c \subseteq b) \Rightarrow c \subseteq d \cap b)$ | UG 4-30 |
| 32. | $(\forall a)(\forall b)(\forall c)((c \subseteq a \wedge c \subseteq b) \Rightarrow c \subseteq a \cap b)$ | UG 3-31 |

□

**Theorem 2.58** $(\forall a)(\forall b)(\forall c)(a \subseteq b \cap c \Rightarrow (a \subseteq b \wedge a \subseteq c))$

***Proof***

| | | |
|---|---|---|
| 1. | $(\forall a)(\forall b)(a \subseteq b \Leftrightarrow (\forall c)(c \in a \Rightarrow c \in b))$ | Ax 2.9 |
| 2. | $(\forall a)(\forall b)(\forall c)(c \in a \cap b \Leftrightarrow (c \in a \wedge c \in b))$ | Ax 2.15 |
| 3. | flag $d$ | FSUG |
| 4. | flag $e$ | FSUG |
| 5. | flag $f$ | FSUG |
| 6. | $d \subseteq e \cap f$ | ACP |
| 7. | $(\forall b)(d \subseteq b \Leftrightarrow (\forall c)(c \in d \Rightarrow c \in b))$ | UI 1 |
| 8. | $d \subseteq e \cap f \Leftrightarrow (\forall c)(c \in d \Rightarrow c \in e \cap f)$ | UI 7 |
| 9. | $(\forall c)(c \in d \Rightarrow c \in e \cap f)$ | BEl 6, 8 |
| 10. | flag $g$ | FSUG |
| 11. | $g \in d \Rightarrow g \in e \cap f$ | UI 9 |
| 12. | $g \in d$ | ACP |
| 13. | $g \in e \cap f$ | MP 11, 12 |
| 14. | $(\forall b)(\forall c)(c \in e \cap b \Leftrightarrow (c \in e \wedge c \in b))$ | UI 2 |
| 15. | $(\forall c)(c \in e \cap f \Leftrightarrow (c \in e \wedge c \in f))$ | UI 14 |
| 16. | $g \in e \cap f \Leftrightarrow (g \in e \wedge g \in f)$ | UI 15 |
| 17. | $g \in e \wedge g \in f$ | BEl 13, 16 |
| 18. | $g \in e$ | S 17 |

| | | |
|---|---|---|
| 19. | $g \in d \Rightarrow g \in e$ | CP 12-18 |
| 20. | $(\forall c)(c \in d \Rightarrow c \in e)$ | UG 10-19 |
| 21. | $d \subseteq e \Leftrightarrow (\forall c)(c \in d \Rightarrow c \in e)$ | UI 7 |
| 22. | $d \subseteq e$ | BEI 20, 21 |
| 23. | flag $h$ | FSUG |
| 24. | $h \in d \Rightarrow h \in e \cap f$ | UI 9 |
| 25. | $h \in d$ | ACP |
| 26. | $h \in e \cap f$ | MP 24, 25 |
| 27. | $(\forall b)(\forall c)(c \in e \cap b \Leftrightarrow (c \in e \wedge c \in b))$ | UI 2 |
| 28. | $(\forall c)(c \in e \cap f \Leftrightarrow (c \in e \wedge c \in f))$ | UI 27 |
| 29. | $h \in e \cap f \Leftrightarrow (h \in e \wedge h \in f)$ | UI 28 |
| 30. | $h \in e \wedge h \in f$ | BEI 26, 29 |
| 31. | $h \in f$ | S 30 |
| 32. | $h \in d \Rightarrow h \in f$ | CP 25-31 |
| 33. | $(\forall c)(c \in d \Rightarrow c \in f)$ | UG 23-32 |
| 34. | $d \subseteq f \Leftrightarrow (\forall c)(c \in d \Rightarrow c \in f)$ | UI 7 |
| 35. | $d \subseteq f$ | BEI 33, 34 |
| 36. | $d \subseteq e \wedge d \subseteq f$ | C 22, 35 |
| 37. | $d \subseteq e \cap f \Rightarrow (d \subseteq e \wedge d \subseteq f)$ | CP 6-36 |
| 38. | $(\forall c)(d \subseteq e \cap c \Rightarrow (d \subseteq e \wedge d \subseteq c))$ | UG 5-37 |
| 39. | $(\forall b)(\forall c)(d \subseteq b \cap c \Rightarrow (d \subseteq b \wedge d \subseteq c))$ | UG 4-38 |
| 40. | $(\forall a)(\forall b)(\forall c)(a \subseteq b \cap c \Rightarrow (a \subseteq b \wedge a \subseteq c))$ | UG 3-39 |

□

**Theorem 2.59** $(\forall a)(\forall b)(a \subseteq b \Leftrightarrow a \cap b = a)$

***Proof***

| | | |
|---|---|---|
| 1. | $(\forall a)(\forall b)(a \cap b \subseteq a)$ | T 2.55 |
| 2. | $(\forall a)(a \subseteq a)$ | T 2.25 |
| 3. | $(\forall a)(\forall b)(\forall c)((c \subseteq a \wedge c \subseteq b) \Rightarrow c \subseteq a \cap b)$ | T 2.57 |
| 4. | $(\forall a)(\forall b)(a = b \Leftrightarrow (a \subseteq b \wedge b \subseteq a))$ | T 2.24 |
| 5. | $(\forall a)(\forall b)(\forall c)(a \subseteq b \cap c \Rightarrow (a \subseteq b \wedge a \subseteq c))$ | T 2.58 |
| 6. | flag $d$ | FSUG |
| 7. | flag $e$ | FSUG |
| 8. | $d \subseteq e$ | ACP |
| 9. | $(\forall b)(d \cap b \subseteq d)$ | UI 1 |
| 10. | $d \cap e \subseteq d$ | UI 9 |
| 11. | $d \subseteq d$ | UI 2 |
| 12. | $d \subseteq d \wedge d \subseteq e$ | C 8, 11 |
| 13. | $(\forall b)(\forall c)((c \subseteq d \wedge c \subseteq b) \Rightarrow c \subseteq d \cap b)$ | UI 3 |
| 14. | $(\forall c)((c \subseteq d \wedge c \subseteq e) \Rightarrow c \subseteq d \cap e)$ | UI 13 |

| | | |
|---|---|---|
| 15. | $(d \subseteq d \wedge d \subseteq e) \Rightarrow d \subseteq d \cap e$ | UI 14 |
| 16. | $d \subseteq d \cap e$ | MP 12, 15 |
| 17. | $d \cap e \subseteq d \wedge d \subseteq d \cap e$ | C 10, 16 |
| 18. | $(\forall b)(d \cap e = b \Leftrightarrow (d \cap e \subseteq b \wedge b \subseteq d \cap e))$ | UI 4 |
| 19. | $d \cap e = d \Leftrightarrow (d \cap e \subseteq d \wedge d \subseteq d \cap e)$ | UI 18 |
| 20. | $d \cap e = d$ | BEl 17, 19 |
| 21. | $d \subseteq e \Rightarrow d \cap e = d$ | CP 8-20 |
| 22. | $d \cap e = d$ | ACP |
| 23. | $(\forall b)(d \cap e = b \Leftrightarrow (d \cap e \subseteq b \wedge b \subseteq d \cap e))$ | UI 4 |
| 24. | $d \cap e = d \Leftrightarrow (d \cap e \subseteq d \wedge d \subseteq d \cap e)$ | UI 23 |
| 25. | $d \cap e \subseteq d \wedge d \subseteq d \cap e$ | BEl 22, 24 |
| 26. | $d \subseteq d \cap e$ | S 25 |
| 27. | $(\forall b)(\forall c)(d \subseteq b \cap c \Rightarrow (d \subseteq b \wedge d \subseteq c))$ | UI 5 |
| 28. | $(\forall c)(d \subseteq d \cap c \Rightarrow (d \subseteq d \wedge d \subseteq c))$ | UI 27 |
| 29. | $d \subseteq d \cap e \Rightarrow (d \subseteq d \wedge d \subseteq e)$ | UI 28 |
| 30. | $d \subseteq d \wedge d \subseteq e$ | MP 26, 29 |
| 31. | $d \subseteq e$ | S 30 |
| 32. | $d \cap e = d \Rightarrow d \subseteq e$ | CP 22-31 |
| 33. | $(d \subseteq e \Rightarrow d \cap e = d) \wedge (d \cap e = d \Rightarrow d \subseteq e)$ | C 21, 32 |
| 34. | $d \subseteq e \Leftrightarrow d \cap e = d$ | BE 33 |
| 35. | $(\forall b)(d \subseteq b \Leftrightarrow d \cap b = d)$ | UG 7-34 |
| 36. | $(\forall a)(\forall b)(a \subseteq b \Leftrightarrow a \cap b = a)$ | UG 6-35 |

□

**Theorem 2.60** $(\forall a)(\forall b)(\forall c)(a \cup (b \cap c) = (a \cup b) \cap (a \cup c))$

***Proof***

| | | |
|---|---|---|
| 1. | $(\forall a)(\forall b)(\forall c)(c \in a \cup b \Leftrightarrow (c \in a \vee c \in b))$ | T 2.18 |
| 2. | $(\forall a)(\forall b)(\forall c)(c \in a \cap b \Leftrightarrow (c \in a \wedge c \in b))$ | Ax 2.15 |
| 3. | $(\forall a)(\forall b)((\forall c)(c \in a \Leftrightarrow c \in b) \Rightarrow a = b)$ | Ax 2.1 |
| 4. | flag $d$ | FSUG |
| 5. | flag $e$ | FSUG |
| 6. | flag $f$ | FSUG |
| 7. | flag $g$ | FSUG |
| 8. | $g \in d \cup (e \cap f)$ | ACP |
| 9. | $(\forall b)(\forall c)(c \in d \cup b \Leftrightarrow (c \in d \vee c \in b))$ | UI 1 |
| 10. | $(\forall c)(c \in d \cup (e \cap f) \Leftrightarrow (c \in d \vee c \in e \cap f))$ | UI 9 |
| 11. | $g \in d \cup (e \cap f) \Leftrightarrow (g \in d \vee g \in e \cap f)$ | UI 10 |
| 12. | $g \in d \vee g \in e \cap f$ | BEl 8, 11 |
| 13. | $(\forall b)(\forall c)(c \in e \cap b \Leftrightarrow (c \in e \wedge c \in b))$ | UI 2 |
| 14. | $(\forall c)(c \in e \cap f \Leftrightarrow (c \in e \wedge c \in f))$ | UI 13 |

| | | |
|---|---|---|
| 15. | $g \in e \cap f \Leftrightarrow (g \in e \wedge g \in f)$ | UI 14 |
| 16. | $g \in d \vee (g \in e \wedge g \in f)$ | D 12, 15 |
| 17. | $(g \in d \vee g \in e) \wedge (g \in d \vee g \in f)$ | Di 16 |
| 18. | $g \in d \vee g \in e$ | S 17 |
| 19. | $(\forall c)(c \in d \cup e \Leftrightarrow (c \in d \vee c \in e))$ | UI 9 |
| 20. | $g \in d \cup e \Leftrightarrow (g \in d \vee g \in e)$ | UI 19 |
| 21. | $g \in d \cup e$ | BEl 18, 20 |
| 22. | $g \in d \vee g \in f$ | S 17 |
| 23. | $(\forall c)(c \in d \cup f \Leftrightarrow (c \in d \vee c \in f))$ | UI 9 |
| 24. | $g \in d \cup f \Leftrightarrow (g \in d \vee g \in f)$ | UI 23 |
| 25. | $g \in d \cup f$ | BEl 22, 24 |
| 26. | $g \in d \cup e \wedge g \in d \cup f$ | C 21, 25 |
| 27. | $(\forall b)(\forall c)(c \in (d \cup e) \cap b \Leftrightarrow (c \in d \cup e \wedge c \in b))$ | UI 2 |
| 28. | $(\forall c)(c \in (d \cup e) \cap (d \cup f) \Leftrightarrow$ <br> $(c \in d \cup e \wedge c \in d \cup f))$ | UI 27 |
| 29. | $g \in (d \cup e) \cap (d \cup f) \Leftrightarrow (g \in d \cup e \wedge g \in d \cup f)$ | UI 28 |
| 30. | $g \in (d \cup e) \cap (d \cup f)$ | BEl 26, 29 |
| 31. | $g \in d \cup (e \cap f) \Rightarrow g \in (d \cup e) \cap (d \cup f)$ | CP 8-30 |
| 32. | $g \in (d \cup e) \cap (d \cup f)$ | ACP |
| 33. | $(\forall b)(\forall c)(c \in (d \cup e) \cap b \Leftrightarrow (c \in d \cup e \wedge c \in b))$ | UI 2 |
| 34. | $(\forall c)(c \in (d \cup e) \cap (d \cup f) \Leftrightarrow$ <br> $(c \in d \cup e \wedge c \in d \cup f))$ | UI 33 |
| 35. | $g \in (d \cup e) \cap (d \cup f) \Leftrightarrow (g \in d \cup e \wedge g \in d \cup f)$ | UI 34 |
| 36. | $g \in d \cup e \wedge g \in d \cup f$ | BEl 32, 35 |
| 37. | $g \in d \cup e$ | S 36 |
| 38. | $(\forall b)(\forall c)(c \in d \cup b \Leftrightarrow (c \in d \vee c \in b))$ | UI 1 |
| 39. | $(\forall c)(c \in d \cup e \Leftrightarrow (c \in d \vee c \in e))$ | UI 38 |
| 40. | $g \in d \cup e \Leftrightarrow (g \in d \vee g \in e)$ | UI 39 |
| 41. | $g \in d \vee g \in e$ | BEl 37, 40 |
| 42. | $g \in d \cup f$ | S 36 |
| 43. | $(\forall c)(c \in d \cup f \Leftrightarrow (c \in d \vee c \in f))$ | UI 38 |
| 44. | $g \in d \cup f \Leftrightarrow (g \in d \vee g \in f)$ | UI 43 |
| 45. | $g \in d \vee g \in f$ | BEl 42, 44 |
| 46. | $(g \in d \vee g \in e) \wedge (g \in d \vee g \in f)$ | C 41, 45 |
| 47. | $g \in d \vee (g \in e \wedge g \in f)$ | Di 46 |
| 48. | $(\forall b)(\forall c)(c \in e \cap b \Leftrightarrow (c \in e \wedge c \in b))$ | UI 2 |
| 49. | $(\forall c)(c \in e \cap f \Leftrightarrow (c \in e \wedge c \in f))$ | UI 48 |
| 50. | $g \in e \cap f \Leftrightarrow (g \in e \wedge g \in f)$ | UI 49 |
| 51. | $g \in d \vee g \in e \cap f$ | D 47, 50 |
| 52. | $(\forall c)(c \in d \cup (e \cap f) \Leftrightarrow (c \in d \vee c \in e \cap f))$ | UI 38 |
| 53. | $g \in d \cup (e \cap f) \Leftrightarrow (g \in d \vee g \in e \cap f)$ | UI 52 |
| 54. | $g \in d \cup (e \cap f)$ | BEl 51, 53 |
| 55. | $g \in (d \cup e) \cap (d \cup f) \Rightarrow g \in d \cup (e \cap f)$ | CP 32-54 |

| | | |
|---|---|---|
| 56. | $(g \in d \cup (e \cap f) \Rightarrow g \in (d \cup e) \cap (d \cup f)) \wedge$ $(g \in (d \cup e) \cap (d \cup f) \Rightarrow g \in d \cup (e \cap f))$ | C 31, 55 |
| 57. | $g \in d \cup (e \cap f) \Leftrightarrow g \in (d \cup e) \cap (d \cup f)$ | BE 56 |
| 58. | $(\forall c)(c \in d \cup (e \cap f) \Leftrightarrow c \in (d \cup e) \cap (d \cup f))$ | UG 7-57 |
| 59. | $(\forall b)((\forall c)(c \in d \cup (e \cap f) \Leftrightarrow c \in b)$ $\Rightarrow d \cup (e \cap f) = b)$ | UI 3 |
| 60. | $(\forall c)(c \in d \cup (e \cap f) \Leftrightarrow c \in (d \cup e) \cap (d \cup f)) \Rightarrow$ $d \cup (e \cap f) = (d \cup e) \cap (d \cup f)$ | UI 59 |
| 61. | $d \cup (e \cap f) = (d \cup e) \cap (d \cup f)$ | MP 58, 60 |
| 62. | $(\forall c)(d \cup (e \cap c) = (d \cup e) \cap (d \cup c))$ | UG 6-61 |
| 63. | $(\forall b)(\forall c)(d \cup (b \cap c) = (d \cup b) \cap (d \cup c))$ | UG 5-62 |
| 64. | $(\forall a)(\forall b)(\forall c)(a \cup (b \cap c) = (a \cup b) \cap (a \cup c))$ | UG 4-63 |

□

**Theorem 2.61** $(\forall a)(\forall b)(\forall c)(a \cap (b \cup c) = (a \cap b) \cup (a \cap c))$

***Proof***

| | | |
|---|---|---|
| 1. | $(\forall a)(\forall b)(\forall c)(c \in a \cap b \Leftrightarrow (c \in a \wedge c \in b))$ | Ax 2.15 |
| 2. | $(\forall a)(\forall b)(\forall c)(c \in a \cup b \Leftrightarrow (c \in a \vee c \in b))$ | T 2.18 |
| 3. | $(\forall a)(\forall b)((\forall c)(c \in a \Leftrightarrow c \in b) \Rightarrow a = b)$ | Ax 2.1 |
| 4. | flag $d$ | FSUG |
| 5. | flag $e$ | FSUG |
| 6. | flag $f$ | FSUG |
| 7. | flag $g$ | FSUG |
| 8. | $g \in d \cap (e \cup f)$ | ACP |
| 9. | $(\forall b)(\forall c)(c \in d \cap b \Leftrightarrow (c \in d \wedge c \in b))$ | UI 1 |
| 10. | $(\forall c)(c \in d \cap (e \cup f) \Leftrightarrow (c \in d \wedge c \in e \cup f))$ | UI 9 |
| 11. | $g \in d \cap (e \cup f) \Leftrightarrow (g \in d \wedge g \in e \cup f)$ | UI 10 |
| 12. | $g \in d \wedge g \in e \cup f$ | BEl 8, 11 |
| 13. | $g \in d$ | S 12 |
| 14. | $g \in e \cup f$ | S 12 |
| 15. | $(\forall b)(\forall c)(c \in e \cup b \Leftrightarrow (c \in e \vee c \in b))$ | UI 2 |
| 16. | $(\forall c)(c \in e \cup f \Leftrightarrow (c \in e \vee c \in f))$ | UI 15 |
| 17. | $g \in e \cup f \Leftrightarrow (g \in e \vee g \in f)$ | UI 16 |
| 18. | $g \in e \vee g \in f$ | BEl 14, 17 |
| 19. | $g \in d \wedge (g \in e \vee g \in f)$ | C 13, 18 |
| 20. | $(g \in d \wedge g \in e) \vee (g \in d \wedge g \in f)$ | Di 19 |
| 21. | $(\forall c)(c \in d \cap e \Leftrightarrow (c \in d \wedge c \in e))$ | UI 9 |
| 22. | $g \in d \cap e \Leftrightarrow (g \in d \wedge g \in e)$ | UI 21 |
| 23. | $(\forall c)(c \in d \cap f \Leftrightarrow (c \in d \wedge c \in f))$ | UI 9 |
| 24. | $g \in d \cap f \Leftrightarrow (g \in d \wedge g \in f)$ | UI 23 |
| 25. | $g \in d \cap e \vee g \in d \cap f$ | D 20, 22, 24 |

| | | |
|---|---|---|
| 26. | $(\forall b)(\forall c)(c \in (d \cap e) \cup b \Leftrightarrow (c \in d \cap e \vee c \in b))$ | UI 2 |
| 27. | $(\forall c)(c \in (d \cap e) \cup (d \cap f) \Leftrightarrow$ $(c \in d \cap e \vee c \in d \cap f))$ | UI 26 |
| 28. | $g \in (d \cap e) \cup (d \cap f) \Leftrightarrow (g \in d \cap e \vee g \in d \cap f)$ | UI 27 |
| 29. | $g \in (d \cap e) \cup (d \cap f)$ | BEl 25, 28 |
| 30. | $g \in d \cap (e \cup f) \Rightarrow g \in (d \cap e) \cup (d \cap f)$ | CP 8-29 |
| 31. | $g \in (d \cap e) \cup (d \cap f)$ | ACP |
| 32. | $(\forall b)(\forall c)(c \in (d \cap e) \cup b \Leftrightarrow (c \in d \cap e \vee c \in b))$ | UI 2 |
| 33. | $(\forall c)(c \in (d \cap e) \cup (d \cap f) \Leftrightarrow$ $(c \in d \cap e \vee c \in d \cap f))$ | UI 32 |
| 34. | $g \in (d \cap e) \cup (d \cap f) \Leftrightarrow (g \in d \cap e \vee g \in d \cap f)$ | UI 33 |
| 35. | $g \in d \cap e \vee g \in d \cap f$ | BEl 31, 34 |
| 36. | $(\forall b)(\forall c)(c \in d \cap b \Leftrightarrow (c \in d \wedge c \in b))$ | UI 1 |
| 37. | $(\forall c)(c \in d \cap e \Leftrightarrow (c \in d \wedge c \in e))$ | UI 36 |
| 38. | $g \in d \cap e \Leftrightarrow (g \in d \wedge g \in e)$ | UI 37 |
| 39. | $(\forall c)(c \in d \cap f \Leftrightarrow (c \in d \wedge c \in f))$ | UI 36 |
| 40. | $g \in d \cap f \Leftrightarrow (g \in d \wedge g \in f)$ | UI 39 |
| 41. | $(g \in d \wedge g \in e) \vee (g \in d \wedge g \in f)$ | D 35, 38, 40 |
| 42. | $g \in d \wedge (g \in e \vee g \in f)$ | Di 41 |
| 43. | $g \in d$ | S 42 |
| 44. | $g \in e \vee g \in f$ | S 42 |
| 45. | $(\forall b)(\forall c)(c \in e \cup b \Leftrightarrow (c \in e \vee c \in b))$ | UI 2 |
| 46. | $(\forall c)(c \in e \cup f \Leftrightarrow (c \in e \vee c \in f))$ | UI 45 |
| 47. | $g \in e \cup f \Leftrightarrow (g \in e \vee g \in f)$ | UI 46 |
| 48. | $g \in e \cup f$ | BEl 44, 47 |
| 49. | $g \in d \wedge g \in e \cup f$ | C 43, 48 |
| 50. | $(\forall c)(c \in d \cap (e \cup f) \Leftrightarrow (c \in d \wedge c \in e \cup f))$ | UI 36 |
| 51. | $g \in d \cap (e \cup f) \Leftrightarrow (g \in d \wedge g \in e \cup f)$ | UI 50 |
| 52. | $g \in d \cap (e \cup f)$ | BEl 49, 51 |
| 53. | $g \in (d \cap e) \cup (d \cap f) \Rightarrow g \in d \cap (e \cup f)$ | CP 31-52 |
| 54. | $(g \in d \cap (e \cup f) \Rightarrow g \in (d \cap e) \cup (d \cap f)) \wedge$ $(g \in (d \cap e) \cup (d \cap f) \Rightarrow g \in d \cap (e \cup f))$ | C 30, 53 |
| 55. | $g \in d \cap (e \cup f) \Leftrightarrow g \in (d \cap e) \cup (d \cap f)$ | BE 54 |
| 56. | $(\forall c)(c \in d \cap (e \cup f) \Leftrightarrow c \in (d \cap e) \cup (d \cap f))$ | UG 7-55 |
| 57. | $(\forall b)((\forall c)(c \in d \cap (e \cup f) \Leftrightarrow c \in b) \Rightarrow$ $d \cap (e \cup f) = b)$ | UI 3 |
| 58. | $(\forall c)(c \in d \cap (e \cup f) \Leftrightarrow c \in (d \cap e) \cup (d \cap f)) \Rightarrow$ $d \cap (e \cup f) = (d \cap e) \cup (d \cap f)$ | UI 57 |
| 59. | $d \cap (e \cup f) = (d \cap e) \cup (d \cap f)$ | MP 56, 58 |
| 60. | $(\forall c)(d \cap (e \cup c) = (d \cap e) \cup (d \cap c))$ | UG 6-59 |
| 61. | $(\forall b)(\forall c)(d \cap (b \cup c) = (d \cap b) \cup (d \cap c))$ | UG 5-60 |
| 62. | $(\forall a)(\forall b)(\forall c)(a \cap (b \cup c) = (a \cap b) \cup (a \cap c))$ | UG 4-61 |

□

**Theorem 2.62** $(\forall a)(\forall b)(a \cap (a \cup b) = a)$

***Proof***

| | | |
|---|---|---|
| 1. | $(\forall a)(\forall b)(a \subseteq a \cup b)$ | T 2.27 |
| 2. | $(\forall a)(\forall b)(a \subseteq b \Leftrightarrow a \cap b = a)$ | T 2.59 |
| 3. | flag $c$ | FSUG |
| 4. | flag $d$ | FSUG |
| 5. | $(\forall b)(c \subseteq c \cup b)$ | UI 1 |
| 6. | $c \subseteq c \cup d$ | UI 5 |
| 7. | $(\forall b)(c \subseteq b \Leftrightarrow c \cap b = c)$ | UI 2 |
| 8. | $c \subseteq c \cup d \Leftrightarrow c \cap (c \cup d) = c$ | UI 7 |
| 9. | $c \cap (c \cup d) = c$ | BEl 6, 8 |
| 10. | $(\forall b)(c \cap (c \cup b) = c)$ | UG 4-9 |
| 11. | $(\forall a)(\forall b)(a \cap (a \cup b) = a)$ | UG 3-10 |

□

**Theorem 2.63** $(\forall a)(\forall b)(a \cup (a \cap b) = a)$

***Proof***

| | | |
|---|---|---|
| 1. | $(\forall a)(\forall b)(a \cap b \subseteq a)$ | T 2.55 |
| 2. | $(\forall a)(\forall b)(a \subseteq b \Leftrightarrow a \cup b = b)$ | T 2.39 |
| 3. | $(\forall a)(\forall b)(a \cup b = b \cup a)$ | T 2.19 |
| 4. | flag $c$ | FSUG |
| 5. | flag $d$ | FSUG |
| 6. | $(\forall b)(c \cap b \subseteq c)$ | UI 1 |
| 7. | $c \cap d \subseteq c$ | UI 6 |
| 8. | $(\forall b)(c \cap d \subseteq b \Leftrightarrow (c \cap d) \cup b = b)$ | UI 2 |
| 9. | $c \cap d \subseteq c \Leftrightarrow (c \cap d) \cup c = c$ | UI 8 |
| 10. | $(c \cap d) \cup c = c$ | BEl 7, 9 |
| 11. | $(\forall b)(c \cup b = b \cup c)$ | UI 3 |
| 12. | $c \cup (c \cap d) = (c \cap d) \cup c$ | UI 11 |
| 13. | $c \cup (c \cap d) = c$ | IT 10, 12 |
| 14. | $(\forall b)(c \cup (c \cap b) = c)$ | UG 5-13 |
| 15. | $(\forall a)(\forall b)(a \cup (a \cap b) = a)$ | UG 4-14 |

□

**Theorem 2.64** $(\forall a)(a \cap \phi = \phi)$

***Proof***

| | | |
|---|---|---|
| 1. | $(\forall a)(\forall b)(a \cap b \subseteq b)$ | T 2.56 |
| 2. | $(\forall a)(\phi \subseteq a)$ | T 2.42 |
| 3. | $(\forall a)(\forall b)(a = b \Leftrightarrow (a \subseteq b \wedge b \subseteq a))$ | T 2.24 |
| 4. | flag $c$ | FSUG |
| 5. | $(\forall b)(c \cap b \subseteq b)$ | UI 1 |
| 6. | $c \cap \phi \subseteq \phi$ | UI 5 |
| 7. | $\phi \subseteq c \cap \phi$ | UI 2 |
| 8. | $c \cap \phi \subseteq \phi \wedge \phi \subseteq c \cap \phi$ | C 6, 7 |
| 9. | $(\forall b)(c \cap \phi = b \Leftrightarrow (c \cap \phi \subseteq b \wedge b \subseteq c \cap \phi))$ | UI 3 |
| 10. | $c \cap \phi = \phi \Leftrightarrow (c \cap \phi \subseteq \phi \wedge \phi \subseteq c \cap \phi)$ | UI 9 |
| 11. | $c \cap \phi = \phi$ | BEl 8, 10 |
| 12. | $(\forall a)(a \cap \phi = \phi)$ | UG 4-11 |

□

**Theorem 2.65** $(\forall a)(a \neq \phi \Rightarrow (\exists b)(\forall c)(c \in b \Leftrightarrow (\forall d)(d \in a \Rightarrow c \in d)))$

***Proof***

| | | |
|---|---|---|
| 1. | $(\forall a)(a \neq \phi \Rightarrow (\exists b)(b \in a))$ | T 2.46 |
| 2. | $(\forall a)(\forall b)(\exists c)(\forall d)(d \in c \Leftrightarrow (d \in a \wedge (\forall e)(e \in b \Rightarrow d \in e)))$ | AS 2.1 |
| 3. | flag $f$ | FSUG |
| 4. | $f \neq \phi$ | ACP |
| 5. | $f \neq \phi \Rightarrow (\exists b)(b \in f)$ | UI 1 |
| 6. | $(\exists b)(b \in f)$ | MP 4, 5 |
| 7. | $g \in f$ | EI 6, flag $g$ |
| 8. | $(\forall b)(\exists c)(\forall d)(d \in c \Leftrightarrow (d \in g \wedge (\forall e)(e \in b \Rightarrow d \in e)))$ | UI 2 |
| 9. | $(\exists c)(\forall d)(d \in c \Leftrightarrow (d \in g \wedge (\forall e)(e \in f \Rightarrow d \in e)))$ | UI 8 |
| 10. | $(\forall d)(d \in h \Leftrightarrow (d \in g \wedge (\forall e)(e \in f \Rightarrow d \in e)))$ | EI 9, flag $h$ |
| 11. | flag $i$ | FSUG |
| 12. | $i \in h \Leftrightarrow (i \in g \wedge (\forall e)(e \in f \Rightarrow i \in e))$ | UI 10 |
| 13. | $(i \in h \Rightarrow (i \in g \wedge (\forall e)(e \in f \Rightarrow i \in e)))\wedge$ $((i \in g \wedge (\forall e)(e \in f \Rightarrow i \in e)) \Rightarrow i \in h)$ | BE 12 |
| 14. | $i \in h \Rightarrow (i \in g \wedge (\forall e)(e \in f \Rightarrow i \in e))$ | S 13 |
| 15. | $(i \in g \wedge (\forall e)(e \in f \Rightarrow i \in e)) \Rightarrow i \in h$ | S 13 |
| 16. | $i \in h$ | ACP |
| 17. | $i \in g \wedge (\forall e)(e \in f \Rightarrow i \in e)$ | MP 14, 16 |
| 18. | $(\forall e)(e \in f \Rightarrow i \in e)$ | S 17 |
| 19. | flag $j$ | FSUG |
| 20. | $j \in f \Rightarrow i \in j$ | UI 18 |
| 21. | $(\forall d)(d \in f \Rightarrow i \in d)$ | UG 19-20 |

| | | |
|---|---|---|
| 22. | $i \in h \Rightarrow (\forall d)(d \in f \Rightarrow i \in d)$ | CP 16-21 |
| 23. | $(\forall d)(d \in f \Rightarrow i \in d)$ | ACP |
| 24. | $g \in f \Rightarrow i \in g$ | UI 23 |
| 25. | $i \in g$ | MP 7, 24 |
| 26. | flag $k$ | FSUG |
| 27. | $k \in f \Rightarrow i \in k$ | UI 23 |
| 28. | $(\forall e)(e \in f \Rightarrow i \in e)$ | UG 26-27 |
| 29. | $i \in g \wedge (\forall e)(e \in f \Rightarrow i \in e)$ | C 25, 28 |
| 30. | $i \in h$ | MP 15, 29 |
| 31. | $(\forall d)(d \in f \Rightarrow i \in d) \Rightarrow i \in h$ | CP 23-30 |
| 32. | $(i \in h \Rightarrow (\forall d)(d \in f \Rightarrow i \in d)) \wedge$ $((\forall d)(d \in f \Rightarrow i \in d) \Rightarrow i \in h)$ | C 22, 31 |
| 33. | $i \in h \Leftrightarrow (\forall d)(d \in f \Rightarrow i \in d)$ | BE 32 |
| 34. | $(\forall c)(c \in h \Leftrightarrow (\forall d)(d \in f \Rightarrow c \in d))$ | UG 11-33 |
| 35. | $(\exists b)(\forall c)(c \in b \Leftrightarrow (\forall d)(d \in f \Rightarrow c \in d))$ | EG 34 |
| 36. | $f \neq \phi \Rightarrow (\exists b)(\forall c)(c \in b \Leftrightarrow (\forall d)(d \in f \Rightarrow c \in d))$ | CP 4-35 |
| 37. | $(\forall a)(a \neq \phi \Rightarrow (\exists b)(\forall c)(c \in b \Leftrightarrow (\forall d)(d \in a \Rightarrow c \in d)))$ | UG 3-36 |

□

**Theorem 2.66** $(\forall a)(a \neq \phi \Rightarrow (\exists! b)(\forall c)(c \in b \Leftrightarrow (\forall d)(d \in a \Rightarrow c \in d)))$

***Proof***

| | | |
|---|---|---|
| 1. | $(\forall a)(a \neq \phi \Rightarrow (\exists b)(\forall c)(c \in b \Leftrightarrow (\forall d)(d \in a \Rightarrow c \in d)))$ | T 2.65 |
| 2. | $(\forall a)(\forall b)((\forall c)(c \in a \Leftrightarrow c \in b) \Rightarrow a = b)$ | Ax 2.1 |
| 3. | flag $e$ | FSUG |
| 4. | $e \neq \phi$ | ACP |
| 5. | $e \neq \phi \Rightarrow (\exists b)(\forall c)(c \in b \Leftrightarrow (\forall d)(d \in e \Rightarrow c \in d))$ | UI 1 |
| 6. | $(\exists b)(\forall c)(c \in b \Leftrightarrow (\forall d)(d \in e \Rightarrow c \in d))$ | MP 4, 5 |
| 7. | $(\forall c)(c \in f \Leftrightarrow (\forall d)(d \in e \Rightarrow c \in d))$ | EI 6, flag $f$ |
| 8. | flag $g$ | FSUG |
| 9. | $(\forall c)(c \in g \Leftrightarrow (\forall d)(d \in e \Rightarrow c \in d))$ | ACP |
| 10. | flag $h$ | FSUG |
| 11. | $h \in g \Leftrightarrow (\forall d)(d \in e \Rightarrow h \in d)$ | UI 9 |
| 12. | $h \in f \Leftrightarrow (\forall d)(d \in e \Rightarrow h \in d)$ | UI 7 |
| 13. | $h \in g \Leftrightarrow h \in f$ | BT 11, 12 |
| 14. | $(\forall c)(c \in g \Leftrightarrow c \in f)$ | UG 10-13 |
| 15. | $(\forall b)((\forall c)(c \in g \Leftrightarrow c \in b) \Rightarrow g = b)$ | UI 2 |
| 16. | $(\forall c)(c \in g \Leftrightarrow c \in f) \Rightarrow g = f$ | UI 15 |
| 17. | $g = f$ | MP 14, 16 |
| 18. | $(\forall c)(c \in g \Leftrightarrow (\forall d)(d \in e \Rightarrow c \in d)) \Rightarrow g = f$ | CP 9-17 |

| | | |
|---|---|---|
| 19. | $(\forall i)((\forall c)(c \in i \Leftrightarrow (\forall d)(d \in e \Rightarrow c \in d)) \Rightarrow i = f)$ | UG 8-18 |
| 20. | $(\forall c)(c \in f \Leftrightarrow (\forall d)(d \in e \Rightarrow c \in d)) \wedge$ $(\forall i)((\forall c)(c \in i \Leftrightarrow (\forall d)(d \in e \Rightarrow c \in d)) \Rightarrow i = f)$ | C 7, 19 |
| 21. | $(\exists b)((\forall c)(c \in b \Leftrightarrow (\forall d)(d \in e \Rightarrow c \in d)) \wedge$ $(\forall i)((\forall c)(c \in i \Leftrightarrow (\forall d)(d \in e \Rightarrow c \in d)) \Rightarrow i = b))$ | EG 20 |
| 22. | $(\exists! b)(\forall c)(c \in b \Leftrightarrow (\forall d)(d \in e \Rightarrow c \in d))$ | UQ 21 |
| 23. | $e \neq \phi \Rightarrow (\exists! b)(\forall c)(c \in b \Leftrightarrow (\forall d)(d \in e \Rightarrow c \in d))$ | CP 4-22 |
| 24. | $(\forall a)(a \neq \phi \Rightarrow (\exists! b)(\forall c)(c \in b \Leftrightarrow (\forall d)(d \in a \Rightarrow c \in d)))$ | UG 3-23 |

□

**Axiom 2.16 (Definitions of Operations, from Theorem 2.66, Which If Applied to the Argument $\alpha$ Give the Terms "$\cap\alpha$" and "intersection of elements of $\alpha$")**

$(\forall a)(a \neq \phi \Rightarrow (\forall b)(b \in \cap a \Leftrightarrow (\forall c)(c \in a \Rightarrow b \in c)))$
$(\forall a)(a \neq \phi \Rightarrow (\forall b)(b \in \text{intersection of elements of } a \Leftrightarrow (\forall c)(c \in a \Rightarrow b \in c)))$

**Theorem 2.67** $(\forall a)(\forall b)(b \in a \Rightarrow \cap a \subseteq b)$

***Proof***

| | | |
|---|---|---|
| 1. | $(\forall a)((\exists b)(b \in a) \Rightarrow a \neq \phi)$ | T 2.48 |
| 2. | $(\forall a)(a \neq \phi \Rightarrow (\forall b)(b \in \cap a \Leftrightarrow (\forall c)(c \in a \Rightarrow b \in c)))$ | Ax 2.16 |
| 3. | $(\forall a)(\forall b)(a \subseteq b \Leftrightarrow (\forall c)(c \in a \Rightarrow c \in b))$ | Ax 2.9 |
| 4. | flag $d$ | FSUG |
| 5. | flag $e$ | FSUG |
| 6. | $e \in d$ | ACP |
| 7. | $(\exists b)(b \in d)$ | EG 6 |
| 8. | $(\exists b)(b \in d) \Rightarrow d \neq \phi$ | UI 1 |
| 9. | $d \neq \phi$ | MP 7, 8 |
| 10. | $d \neq \phi \Rightarrow (\forall b)(b \in \cap d \Leftrightarrow (\forall c)(c \in d \Rightarrow b \in c))$ | UI 2 |
| 11. | $(\forall b)(b \in \cap d \Leftrightarrow (\forall c)(c \in d \Rightarrow b \in c))$ | MP 9, 10 |
| 12. | flag $f$ | FSUG |
| 13. | $f \in \cap d \Leftrightarrow (\forall c)(c \in d \Rightarrow f \in c)$ | UI 11 |
| 14. | $f \in \cap d$ | ACP |
| 15. | $(\forall c)(c \in d \Rightarrow f \in c)$ | BEI 13, 14 |
| 16. | $e \in d \Rightarrow f \in e$ | UI 15 |
| 17. | $f \in e$ | MP 6, 16 |
| 18. | $f \in \cap d \Rightarrow f \in e$ | CP 14-17 |
| 19. | $(\forall c)(c \in \cap d \Rightarrow c \in e)$ | UG 12-18 |
| 20. | $(\forall b)(\cap d \subseteq b \Leftrightarrow (\forall c)(c \in \cap d \Rightarrow c \in b))$ | UI 3 |
| 21. | $\cap d \subseteq e \Leftrightarrow (\forall c)(c \in \cap d \Rightarrow c \in e)$ | UI 20 |
| 22. | $\cap d \subseteq e$ | BEI 19, 21 |

| | | |
|---|---|---|
| 23. | $e \in d \Rightarrow \cap d \subseteq e$ | CP 6-22 |
| 24. | $(\forall b)(b \in d \Rightarrow \cap d \subseteq b)$ | UG 5-23 |
| 25. | $(\forall a)(\forall b)(b \in a \Rightarrow \cap a \subseteq b)$ | UG 4-24 |

□

**Theorem 2.68** $(\forall a)(\forall b)(\cap\{a, b\} = a \cap b)$

***Proof***

| | | |
|---|---|---|
| 1. | $(\forall a)(\forall b)(a \in \{a, b\})$ | T 2.5 |
| 2. | $(\forall a)((\exists b)(b \in a) \Rightarrow a \neq \phi)$ | T 2.48 |
| 3. | $(\forall a)(a \neq \phi \Rightarrow (\forall b)(b \in \cap a \Leftrightarrow (\forall c)(c \in a \Rightarrow b \in c)))$ | Ax 2.16 |
| 4. | $(\forall a)(\forall b)(a \in \{b, a\})$ | T 2.7 |
| 5. | $(\forall a)(\forall b)(\forall c)(c \in a \cap b \Leftrightarrow (c \in a \wedge c \in b))$ | Ax 2.15 |
| 6. | $(\forall a)(\forall b)(\forall c)(c \in \{a, b\} \Leftrightarrow (c = a \vee c = b))$ | Ax 2.3 |
| 7. | $(\forall a)(\forall b)((\forall c)(c \in a \Leftrightarrow c \in b) \Rightarrow a = b)$ | Ax 2.1 |
| 8. | flag $d$ | FSUG |
| 9. | flag $e$ | FSUG |
| 10. | $(\forall b)(d \in \{d, b\})$ | UI 1 |
| 11. | $d \in \{d, e\}$ | UI 10 |
| 12. | $(\exists b)(b \in \{d, e\})$ | EG 11 |
| 13. | $(\exists b)(b \in \{d, e\}) \Rightarrow \{d, e\} \neq \phi$ | UI 2 |
| 14. | $\{d, e\} \neq \phi$ | MP 12, 13 |
| 15. | $\{d, e\} \neq \phi \Rightarrow$ $(\forall b)(b \in \cap\{d, e\} \Leftrightarrow (\forall c)(c \in \{d, e\} \Rightarrow b \in c))$ | UI 3 |
| 16. | $(\forall b)(b \in \cap\{d, e\} \Leftrightarrow (\forall c)(c \in \{d, e\} \Rightarrow b \in c))$ | MP 14, 15 |
| 17. | flag $f$ | FSUG |
| 18. | $f \in \cap\{d, e\}$ | ACP |
| 19. | $f \in \cap\{d, e\} \Leftrightarrow (\forall c)(c \in \{d, e\} \Rightarrow f \in c)$ | UI 16 |
| 20. | $(\forall c)(c \in \{d, e\} \Rightarrow f \in c)$ | BEI 18, 19 |
| 21. | $d \in \{d, e\} \Rightarrow f \in d$ | UI 20 |
| 22. | $f \in d$ | MP 11, 21 |
| 23. | $(\forall b)(e \in \{b, e\})$ | UI 4 |
| 24. | $e \in \{d, e\}$ | UI 23 |
| 25. | $e \in \{d, e\} \Rightarrow f \in e$ | UI 20 |
| 26. | $f \in e$ | MP 24, 25 |
| 27. | $f \in d \wedge f \in e$ | C 22, 26 |
| 28. | $(\forall b)(\forall c)(c \in d \cap b \Leftrightarrow (c \in d \wedge c \in b))$ | UI 5 |
| 29. | $(\forall c)(c \in d \cap e \Leftrightarrow (c \in d \wedge c \in e))$ | UI 28 |
| 30. | $f \in d \cap e \Leftrightarrow (f \in d \wedge f \in e)$ | UI 29 |
| 31. | $f \in d \cap e$ | BEI 27, 30 |
| 32. | $f \in \cap\{d, e\} \Rightarrow f \in d \cap e$ | CP 18-31 |

| | | |
|---|---|---|
| 33. | $f \in d \cap e$ | ACP |
| 34. | $(\forall b)(\forall c)(c \in d \cap b \Leftrightarrow (c \in d \wedge c \in b))$ | UI 5 |
| 35. | $(\forall c)(c \in d \cap e \Leftrightarrow (c \in d \wedge c \in e))$ | UI 34 |
| 36. | $f \in d \cap e \Leftrightarrow (f \in d \wedge f \in e)$ | UI 35 |
| 37. | $f \in d \wedge f \in e$ | BEl 33, 36 |
| 38. | $f \in d$ | S 37 |
| 39. | $f \in e$ | S 37 |
| 40. | flag $g$ | FSUG |
| 41. | $g \in \{d, e\}$ | ACP |
| 42. | $(\forall b)(\forall c)(c \in \{d, b\} \Leftrightarrow (c = d \vee c = b))$ | UI 6 |
| 43. | $(\forall c)(c \in \{d, e\} \Leftrightarrow (c = d \vee c = e))$ | UI 42 |
| 44. | $g \in \{d, e\} \Leftrightarrow (g = d \vee g = e)$ | UI 43 |
| 45. | $g = d \vee g = e$ | BEl 41, 44 |
| 46. | $g = d$ | ACP |
| 47. | $f \in g$ | IS 38, 46 |
| 48. | $g = d \Rightarrow f \in g$ | CP 46-47 |
| 49. | $g = e$ | ACP |
| 50. | $f \in g$ | IS 39, 49 |
| 51. | $g = e \Rightarrow f \in g$ | CP 49-50 |
| 52. | $f \in g \vee f \in g$ | D 45, 48, 51 |
| 53. | $f \in g$ | Du 52 |
| 54. | $g \in \{d, e\} \Rightarrow f \in g$ | CP 41-53 |
| 55. | $(\forall c)(c \in \{d, e\} \Rightarrow f \in c)$ | UG 40-54 |
| 56. | $f \in \cap\{d, e\} \Leftrightarrow (\forall c)(c \in \{d, e\} \Rightarrow f \in c)$ | UI 16 |
| 57. | $f \in \cap\{d, e\}$ | BEl 55, 56 |
| 58. | $f \in d \cap e \Rightarrow f \in \cap\{d, e\}$ | CP 33-57 |
| 59. | $(f \in \cap\{d, e\} \Rightarrow f \in d \cap e) \wedge$ $(f \in d \cap e \Rightarrow f \in \cap\{d, e\})$ | C 32, 58 |
| 60. | $f \in \cap\{d, e\} \Leftrightarrow f \in d \cap e$ | BE 59 |
| 61. | $(\forall c)(c \in \cap\{d, e\} \Leftrightarrow c \in d \cap e)$ | UG 17-60 |
| 62. | $(\forall b)((\forall c)(c \in \cap\{d, e\} \Leftrightarrow c \in b) \Rightarrow \cap\{d, e\} = b)$ | UI 7 |
| 63. | $(\forall c)(c \in \cap\{d, e\} \Leftrightarrow c \in d \cap e) \Rightarrow \cap\{d, e\} = d \cap e$ | UI 62 |
| 64. | $\cap \{d, e\} = d \cap e$ | MP 61, 63 |
| 65. | $(\forall b)(\cap\{d, b\} = d \cap b)$ | UG 9-64 |
| 66. | $(\forall a)(\forall b)(\cap\{a, b\} = a \cap b)$ | UG 8-65 |

□

## 2.6 Russell's Paradox and Non-existence of Universal Set

**Theorem 2.69 (Russell's Paradox)** $\neg(\exists a)(\forall b)(b \in a \Leftrightarrow b \notin b)$

***Proof***

| | | |
|---|---|---|
| 1. | $(\forall a)(\forall b)(a \notin b \Leftrightarrow \neg(a \in b))$ | Ax 2.12 |
| 2. | $(\exists a)(\forall b)(b \in a \Leftrightarrow b \notin b)$ | AIP |
| 3. | $(\forall b)(b \in c \Leftrightarrow b \notin b)$ | EI 2, flag $c$ |
| 4. | $c \in c \Leftrightarrow c \notin c$ | UI 3 |
| 5. | $(\forall b)(c \notin b \Leftrightarrow \neg(c \in b))$ | UI 1 |
| 6. | $c \notin c \Leftrightarrow \neg(c \in c)$ | UI 5 |
| 7. | $c \in c \Leftrightarrow \neg(c \in c)$ | BT 4, 6 |
| 8. | $(c \in c \Rightarrow \neg(c \in c)) \wedge ((\neg(c \in c)) \Rightarrow c \in c)$ | BE 7 |
| 9. | $c \in c \Rightarrow \neg(c \in c)$ | S 8 |
| 10. | $(\neg(c \in c)) \vee \neg(c \in c)$ | CE 9 |
| 11. | $\neg(c \in c)$ | Du 10 |
| 12. | $(\neg(c \in c)) \Rightarrow c \in c$ | S 8 |
| 13. | $c \in c$ | MP 11, 12 |
| 14. | $c \in c \wedge \neg(c \in c)$ | C 11, 13 |
| 15. | $\neg(\exists a)(\forall b)(b \in a \Leftrightarrow b \notin b)$ | IP 2-14 |

□

**Theorem 2.70 (Non-existence of Universal Set)** $\neg(\exists a)(\forall b)(b \in a)$

***Proof***

| | | |
|---|---|---|
| 1. | $(\forall a)(\exists b)(\forall c)(c \in b \Leftrightarrow (c \in a \wedge \neg(c \in c)))$ | AS 2.1 |
| 2. | $(\exists a)(\forall b)(b \in a)$ | AIP |
| 3. | $(\forall b)(b \in d)$ | EI 2, flag $d$ |
| 4. | $(\exists b)(\forall c)(c \in b \Leftrightarrow (c \in d \wedge \neg(c \in c)))$ | UI 1 |
| 5. | $(\forall c)(c \in e \Leftrightarrow (c \in d \wedge \neg(c \in c)))$ | EI 4, flag $e$ |
| 6. | $e \in e \Leftrightarrow (e \in d \wedge \neg(e \in e))$ | UI 5 |
| 7. | $(e \in e \Rightarrow (e \in d \wedge \neg(e \in e))) \wedge ((e \in d \wedge \neg(e \in e)) \Rightarrow e \in e)$ | BE 6 |
| 8. | $e \in e \Rightarrow (e \in d \wedge \neg(e \in e))$ | S 7 |
| 9. | $(\neg(e \in e)) \vee (e \in d \wedge \neg(e \in e))$ | CE 8 |
| 10. | $((\neg(e \in e)) \vee e \in d) \wedge ((\neg(e \in e)) \vee \neg(e \in e))$ | Di 9 |
| 11. | $(\neg(e \in e)) \vee \neg(e \in e)$ | S 10 |
| 12. | $\neg(e \in e)$ | Du 11 |
| 13. | $(e \in d \wedge \neg(e \in e)) \Rightarrow e \in e$ | S 7 |
| 14. | $e \in d$ | UI 3 |
| 15. | $e \in d \wedge \neg(e \in e)$ | C 12, 14 |
| 16. | $e \in e$ | MP 13, 15 |
| 17. | $e \in e \wedge \neg(e \in e)$ | C 12, 16 |
| 18. | $\neg(\exists a)(\forall b)(b \in a)$ | IP 2-17 |

□

## 2.7 Difference

**Theorem 2.71** $(\forall a)(\forall b)(\exists! c)(\forall d)(d \in c \Leftrightarrow (d \in a \wedge d \notin b))$

***Proof***

| | | |
|---|---|---|
| 1. | $(\forall a)(\forall b)(\exists c)(\forall d)(d \in c \Leftrightarrow (d \in a \wedge d \notin b))$ | AS 2.1 |
| 2. | $(\forall a)(\forall b)((\forall c)(c \in a \Leftrightarrow c \in b) \Rightarrow a = b)$ | Ax 2.1 |
| 3. | flag $e$ | FSUG |
| 4. | flag $f$ | FSUG |
| 5. | $(\forall b)(\exists c)(\forall d)(d \in c \Leftrightarrow (d \in e \wedge d \notin b))$ | UI 1 |
| 6. | $(\exists c)(\forall d)(d \in c \Leftrightarrow (d \in e \wedge d \notin f))$ | UI 5 |
| 7. | flag $g$ | FSUG |
| 8. | flag $h$ | FSUG |
| 9. | $(\forall d)(d \in g \Leftrightarrow (d \in e \wedge d \notin f)) \wedge$ $(\forall d)(d \in h \Leftrightarrow (d \in e \wedge d \notin f))$ | ACP |
| 10. | $(\forall d)(d \in g \Leftrightarrow (d \in e \wedge d \notin f))$ | S 9 |
| 11. | $(\forall d)(d \in h \Leftrightarrow (d \in e \wedge d \notin f))$ | S 9 |
| 12. | flag $i$ | FSUG |
| 13. | $i \in g \Leftrightarrow (i \in e \wedge i \notin f)$ | UI 10 |
| 14. | $i \in h \Leftrightarrow (i \in e \wedge i \notin f)$ | UI 11 |
| 15. | $i \in g \Leftrightarrow i \in h$ | BT 13, 14 |
| 16. | $(\forall c)(c \in g \Leftrightarrow c \in h)$ | UG 12-15 |
| 17. | $(\forall b)((\forall c)(c \in g \Leftrightarrow c \in b) \Rightarrow g = b)$ | UI 2 |
| 18. | $(\forall c)(c \in g \Leftrightarrow c \in h) \Rightarrow g = h$ | UI 17 |
| 19. | $g = h$ | MP 16, 18 |
| 20. | $((\forall d)(d \in g \Leftrightarrow (d \in e \wedge d \notin f)) \wedge$ $(\forall d)(d \in h \Leftrightarrow (d \in e \wedge d \notin f))) \Rightarrow g = h$ | CP 9-19 |
| 21. | $(\forall j)(((\forall d)(d \in g \Leftrightarrow (d \in e \wedge d \notin f)) \wedge$ $(\forall d)(d \in j \Leftrightarrow (d \in e \wedge d \notin f))) \Rightarrow g = j)$ | UG 8-20 |
| 22. | $(\forall k)(\forall j)(((\forall d)(d \in k \Leftrightarrow (d \in e \wedge d \notin f)) \wedge$ $(\forall d)(d \in j \Leftrightarrow (d \in e \wedge d \notin f))) \Rightarrow k = j)$ | UG 7-21 |
| 23. | $(\exists c)(\forall d)(d \in c \Leftrightarrow (d \in e \wedge d \notin f)) \wedge$ $(\forall k)(\forall j)(((\forall d)(d \in k \Leftrightarrow (d \in e \wedge d \notin f)) \wedge$ $(\forall d)(d \in j \Leftrightarrow (d \in e \wedge d \notin f))) \Rightarrow k = j)$ | C 6, 22 |
| 24. | $(\exists! c)(\forall d)(d \in c \Leftrightarrow (d \in e \wedge d \notin f))$ | UQ 23 |
| 25. | $(\forall b)(\exists! c)(\forall d)(d \in c \Leftrightarrow (d \in e \wedge d \notin b))$ | UG 4-24 |
| 26. | $(\forall a)(\forall b)(\exists! c)(\forall d)(d \in c \Leftrightarrow (d \in a \wedge d \notin b))$ | UG 3-25 |

□

**Axiom 2.17 (Definitions of Operations, from Theorem 2.71, Which If Applied to the Arguments $\alpha$ and $\beta$ Give the Terms "$\alpha\backslash\beta$" and "difference of $\alpha$ and $\beta$")**

$(\forall a)(\forall b)(\forall c)(c \in a\backslash b \Leftrightarrow (c \in a \wedge c \notin b))$
$(\forall a)(\forall b)(\forall c)(c \in \text{difference of } a \text{ and } b \Leftrightarrow (c \in a \wedge c \notin b))$

**Theorem 2.72** $(\forall a)(a\backslash a = \phi)$

***Proof***

| | | |
|---|---|---|
| 1. | $(\forall a)(\forall b)(\forall c)(c \in a\backslash b \Leftrightarrow (c \in a \wedge c \notin b))$ | Ax 2.17 |
| 2. | $(\forall a)(\forall b)(a \notin b \Leftrightarrow \neg(a \in b))$ | Ax 2.12 |
| 3. | $(\forall a)((\forall b)(b \notin a) \Rightarrow a = \phi)$ | T 2.45 |
| 4. | flag $d$ | FSUG |
| 5. | flag $e$ | FSUG |
| 6. | $e \in d\backslash d$ | AIP |
| 7. | $(\forall b)(\forall c)(c \in d\backslash b \Leftrightarrow (c \in d \wedge c \notin b))$ | UI 1 |
| 8. | $(\forall c)(c \in d\backslash d \Leftrightarrow (c \in d \wedge c \notin d))$ | UI 7 |
| 9. | $e \in d\backslash d \Leftrightarrow (e \in d \wedge e \notin d)$ | UI 8 |
| 10. | $e \in d \wedge e \notin d$ | BEl 6, 9 |
| 11. | $e \in d$ | S 10 |
| 12. | $e \notin d$ | S 10 |
| 13. | $(\forall b)(e \notin b \Leftrightarrow \neg(e \in b))$ | UI 2 |
| 14. | $e \notin d \Leftrightarrow \neg(e \in d)$ | UI 13 |
| 15. | $\neg(e \in d)$ | BEl 12, 14 |
| 16. | $e \in d \wedge \neg(e \in d)$ | C 11, 15 |
| 17. | $\neg(e \in d\backslash d)$ | IP 6-16 |
| 18. | $(\forall b)(e \notin b \Leftrightarrow \neg(e \in b))$ | UI 2 |
| 19. | $e \notin d\backslash d \Leftrightarrow \neg(e \in d\backslash d)$ | UI 18 |
| 20. | $e \notin d\backslash d$ | BEl 17, 19 |
| 21. | $(\forall b)(b \notin d\backslash d)$ | UG 5-20 |
| 22. | $(\forall b)(b \notin d\backslash d) \Rightarrow d\backslash d = \phi$ | UI 3 |
| 23. | $d\backslash d = \phi$ | MP 21, 22 |
| 24. | $(\forall a)(a\backslash a = \phi)$ | UG 4-23 |

□

**Theorem 2.73** $(\forall a)(\forall b)(a\backslash(b\backslash a) = a)$

***Proof***

| | | |
|---|---|---|
| 1. | $(\forall a)(\forall b)(\forall c)(c \in a\backslash b \Leftrightarrow (c \in a \wedge c \notin b))$ | Ax 2.17 |
| 2. | $(\forall a)(\forall b)(a \notin b \Leftrightarrow \neg(a \in b))$ | Ax 2.12 |
| 3. | $(\forall a)(\forall b)((\forall c)(c \in a \Leftrightarrow c \in b) \Rightarrow a = b)$ | Ax 2.1 |
| 4. | flag $d$ | FSUG |
| 5. | flag $e$ | FSUG |
| 6. | flag $f$ | FSUG |
| 7. | $f \in d\backslash(e\backslash d)$ | ACP |
| 8. | $(\forall b)(\forall c)(c \in d\backslash b \Leftrightarrow (c \in d \wedge c \notin b))$ | UI 1 |
| 9. | $(\forall c)(c \in d\backslash(e\backslash d) \Leftrightarrow (c \in d \wedge c \notin e\backslash d))$ | UI 8 |
| 10. | $f \in d\backslash(e\backslash d) \Leftrightarrow (f \in d \wedge f \notin e\backslash d)$ | UI 9 |
| 11. | $f \in d \wedge f \notin e\backslash d$ | BEl 7, 10 |
| 12. | $f \in d$ | S 11 |
| 13. | $f \in d\backslash(e\backslash d) \Rightarrow f \in d$ | CP 7-12 |
| 14. | $f \in d$ | ACP |
| 15. | $\neg\neg(f \in d)$ | DN 14 |
| 16. | $(\forall b)(f \notin b \Leftrightarrow \neg(f \in b))$ | UI 2 |
| 17. | $f \notin d \Leftrightarrow \neg(f \in d)$ | UI 16 |
| 18. | $\neg(f \notin d)$ | NBE 15, 17 |
| 19. | $(\neg(f \in e)) \vee \neg(f \notin d)$ | Ad 18 |
| 20. | $\neg(f \in e \wedge f \notin d)$ | DM 19 |
| 21. | $(\forall b)(\forall c)(c \in e\backslash b \Leftrightarrow (c \in e \wedge c \notin b))$ | UI 1 |
| 22. | $(\forall c)(c \in e\backslash d \Leftrightarrow (c \in e \wedge c \notin d))$ | UI 21 |
| 23. | $f \in e\backslash d \Leftrightarrow (f \in e \wedge f \notin d)$ | UI 22 |
| 24. | $\neg(f \in e\backslash d)$ | NBE 20, 23 |
| 25. | $f \notin e\backslash d \Leftrightarrow \neg(f \in e\backslash d)$ | UI 16 |
| 26. | $f \notin e\backslash d$ | BEl 24, 25 |
| 27. | $f \in d \wedge f \notin e\backslash d$ | C 14, 26 |
| 28. | $(\forall b)(\forall c)(c \in d\backslash b \Leftrightarrow (c \in d \wedge c \notin b))$ | UI 1 |
| 29. | $(\forall c)(c \in d\backslash(e\backslash d) \Leftrightarrow (c \in d \wedge c \notin e\backslash d))$ | UI 28 |
| 30. | $f \in d\backslash(e\backslash d) \Leftrightarrow (f \in d \wedge f \notin e\backslash d)$ | UI 29 |
| 31. | $f \in d\backslash(e\backslash d)$ | BEl 27, 30 |
| 32. | $f \in d \Rightarrow f \in d\backslash(e\backslash d)$ | CP 14-31 |
| 33. | $(f \in d\backslash(e\backslash d) \Rightarrow f \in d) \wedge (f \in d \Rightarrow f \in d\backslash(e\backslash d))$ | C 13, 32 |
| 34. | $f \in d\backslash(e\backslash d) \Leftrightarrow f \in d$ | BE 33 |
| 35. | $(\forall c)(c \in d\backslash(e\backslash d) \Leftrightarrow c \in d)$ | UG 6-34 |
| 36. | $(\forall b)((\forall c)(c \in d\backslash(e\backslash d) \Leftrightarrow c \in b) \Rightarrow d\backslash(e\backslash d) = b)$ | UI 3 |
| 37. | $(\forall c)(c \in d\backslash(e\backslash d) \Leftrightarrow c \in d) \Rightarrow d\backslash(e\backslash d) = d$ | UI 36 |
| 38. | $d\backslash(e\backslash d) = d$ | MP 35, 37 |
| 39. | $(\forall b)(d\backslash(b\backslash d) = d)$ | UG 5-38 |
| 40. | $(\forall a)(\forall b)(a\backslash(b\backslash a) = a$ | UG 4-39 |

□

**Theorem 2.74** $(\forall a)(\forall b)(a \backslash b \subseteq a)$

***Proof***

| | | |
|---|---|---|
| 1. | $(\forall a)(\forall b)(\forall c)(c \in a \backslash b \Leftrightarrow (c \in a \wedge c \notin b))$ | Ax 2.17 |
| 2. | $(\forall a)(\forall b)(a \subseteq b \Leftrightarrow (\forall c)(c \in a \Rightarrow c \in b))$ | Ax 2.9 |
| 3. | flag $d$ | FSUG |
| 4. | flag $e$ | FSUG |
| 5. | flag $f$ | FSUG |
| 6. | $f \in d \backslash e$ | ACP |
| 7. | $(\forall b)(\forall c)(c \in d \backslash b \Leftrightarrow (c \in d \wedge c \notin b))$ | UI 1 |
| 8. | $(\forall c)(c \in d \backslash e \Leftrightarrow (c \in d \wedge c \notin e))$ | UI 7 |
| 9. | $f \in d \backslash e \Leftrightarrow (f \in d \wedge f \notin e)$ | UI 8 |
| 10. | $f \in d \wedge f \notin e$ | BEl 6, 9 |
| 11. | $f \in d$ | S 10 |
| 12. | $f \in d \backslash e \Rightarrow f \in d$ | CP 6-11 |
| 13. | $(\forall c)(c \in d \backslash e \Rightarrow c \in d)$ | UG 5-12 |
| 14. | $(\forall b)(d \backslash e \subseteq b \Leftrightarrow (\forall c)(c \in d \backslash e \Rightarrow c \in b))$ | UI 2 |
| 15. | $d \backslash e \subseteq d \Leftrightarrow (\forall c)(c \in d \backslash e \Rightarrow c \in d)$ | UI 14 |
| 16. | $d \backslash e \subseteq d$ | BEl 13, 15 |
| 17. | $(\forall b)(d \backslash b \subseteq d)$ | UG 4-16 |
| 18. | $(\forall a)(\forall b)(a \backslash b \subseteq a)$ | UG 3-17 |

□

**Theorem 2.75** $(\forall a)(a \backslash \phi = a)$

***Proof***

| | | |
|---|---|---|
| 1. | $(\forall a)(\forall b)(\forall c)(c \in a \backslash b \Leftrightarrow (c \in a \wedge c \notin b))$ | Ax 2.17 |
| 2. | $(\forall a)(a \notin \phi)$ | Ax 2.14 |
| 3. | $(\forall a)(\forall b)((\forall c)(c \in a \Leftrightarrow c \in b) \Rightarrow a = b)$ | Ax 2.1 |
| 4. | flag $d$ | FSUG |
| 5. | flag $e$ | FSUG |
| 6. | $e \in d \backslash \phi$ | ACP |
| 7. | $(\forall b)(\forall c)(c \in d \backslash b \Leftrightarrow (c \in d \wedge c \notin b))$ | UI 1 |
| 8. | $(\forall c)(c \in d \backslash \phi \Leftrightarrow (c \in d \wedge c \notin \phi))$ | UI 7 |
| 9. | $e \in d \backslash \phi \Leftrightarrow (e \in d \wedge e \notin \phi)$ | UI 8 |
| 10. | $e \in d \wedge e \notin \phi$ | BEl 6, 9 |
| 11. | $e \in d$ | S 10 |
| 12. | $e \in d \backslash \phi \Rightarrow e \in d$ | CP 6-11 |
| 13. | $e \in d$ | ACP |
| 14. | $e \notin \phi$ | UI 2 |

| | | |
|---|---|---|
| 15. | $e \in d \wedge e \notin \phi$ | C 13, 14 |
| 16. | $(\forall b)(\forall c)(c \in d\backslash b \Leftrightarrow (c \in d \wedge c \notin b))$ | UI 1 |
| 17. | $(\forall c)(c \in d\backslash\phi \Leftrightarrow (c \in d \wedge c \notin \phi))$ | UI 16 |
| 18. | $e \in d\backslash\phi \Leftrightarrow (e \in d \wedge e \notin \phi)$ | UI 17 |
| 19. | $e \in d\backslash\phi$ | BEl 15, 18 |
| 20. | $e \in d \Rightarrow e \in d\backslash\phi$ | CP 13-19 |
| 21. | $(e \in d\backslash\phi \Rightarrow e \in d) \wedge (e \in d \Rightarrow e \in d\backslash\phi)$ | C 12, 20 |
| 22. | $e \in d\backslash\phi \Leftrightarrow e \in d$ | BE 21 |
| 23. | $(\forall c)(c \in d\backslash\phi \Leftrightarrow c \in d)$ | UG 5-22 |
| 24. | $(\forall b)((\forall c)(c \in d\backslash\phi \Leftrightarrow c \in b) \Rightarrow d\backslash\phi = b)$ | UI 3 |
| 25. | $(\forall c)(c \in d\backslash\phi \Leftrightarrow c \in d) \Rightarrow d\backslash\phi = d$ | UI 24 |
| 26. | $d\backslash\phi = d$ | MP 23, 25 |
| 27. | $(\forall a)(a\backslash\phi = a)$ | UG 4-26 |

□

**Theorem 2.76** $(\forall a)(\forall b)(a \subseteq b \Leftrightarrow a\backslash b = \phi)$

***Proof***

| | | |
|---|---|---|
| 1. | $(\forall a)(\forall b)(a \subseteq b \Leftrightarrow (\forall c)(c \in a \Rightarrow c \in b))$ | Ax 2.9 |
| 2. | $(\forall a)(\forall b)(\forall c)(c \in a\backslash b \Leftrightarrow (c \in a \wedge c \notin b))$ | Ax 2.17 |
| 3. | $(\forall a)(\forall b)(a \notin b \Leftrightarrow \neg(a \in b))$ | Ax 2.12 |
| 4. | $(\forall a)((\forall b)(b \notin a) \Rightarrow a = \phi)$ | T 2.45 |
| 5. | $(\forall a)(a = \phi \Rightarrow (\forall b)(b \notin a))$ | T 2.47 |
| 6. | flag $d$ | FSUG |
| 7. | flag $e$ | FSUG |
| 8. | $d \subseteq e$ | ACP |
| 9. | $(\forall b)(d \subseteq b \Leftrightarrow (\forall c)(c \in d \Rightarrow c \in b))$ | UI 1 |
| 10. | $d \subseteq e \Leftrightarrow (\forall c)(c \in d \Rightarrow c \in e)$ | UI 9 |
| 11. | $(\forall c)(c \in d \Rightarrow c \in e)$ | BEl 8, 10 |
| 12. | flag $f$ | FSUG |
| 13. | $f \in d\backslash e$ | AIP |
| 14. | $(\forall b)(\forall c)(c \in d\backslash b \Leftrightarrow (c \in d \wedge c \notin b))$ | UI 2 |
| 15. | $(\forall c)(c \in d\backslash e \Leftrightarrow (c \in d \wedge c \notin e))$ | UI 14 |
| 16. | $f \in d\backslash e \Leftrightarrow (f \in d \wedge f \notin e)$ | UI 15 |
| 17. | $f \in d \wedge f \notin e$ | BEl 13, 16 |
| 18. | $f \in d$ | S 17 |
| 19. | $f \notin e$ | S 17 |
| 20. | $(\forall b)(f \notin b \Leftrightarrow \neg(f \in b))$ | UI 3 |
| 21. | $f \notin e \Leftrightarrow \neg(f \in e)$ | UI 20 |
| 22. | $\neg(f \in e)$ | BEl 19, 21 |
| 23. | $f \in d \Rightarrow f \in e$ | UI 11 |

| | | |
|---|---|---|
| 24. | $f \in e$ | MP 18, 23 |
| 25. | $f \in e \wedge \neg(f \in e)$ | C 22, 24 |
| 26. | $\neg(f \in d\backslash e)$ | IP 13-25 |
| 27. | $(\forall b)(f \notin b \Leftrightarrow \neg(f \in b))$ | UI 3 |
| 28. | $f \notin d\backslash e \Leftrightarrow \neg(f \in d\backslash e)$ | UI 27 |
| 29. | $f \notin d\backslash e$ | BEl 26, 28 |
| 30. | $(\forall b)(b \notin d\backslash e)$ | UG 12-29 |
| 31. | $(\forall b)(b \notin d\backslash e) \Rightarrow d\backslash e = \phi$ | UI 4 |
| 32. | $d\backslash e = \phi$ | MP 30, 31 |
| 33. | $d \subseteq e \Rightarrow d\backslash e = \phi$ | CP 8-32 |
| 34. | $d\backslash e = \phi$ | ACP |
| 35. | $d\backslash e = \phi \Rightarrow (\forall b)(b \notin d\backslash e)$ | UI 5 |
| 36. | $(\forall b)(b \notin d\backslash e)$ | MP 34, 35 |
| 37. | flag $g$ | FSUG |
| 38. | $g \notin d\backslash e$ | UI 36 |
| 39. | $(\forall b)(g \notin b \Leftrightarrow \neg(g \in b))$ | UI 3 |
| 40. | $g \notin d\backslash e \Leftrightarrow \neg(g \in d\backslash e)$ | UI 39 |
| 41. | $\neg(g \in d\backslash e)$ | BEl 38, 40 |
| 42. | $(\forall b)(\forall c)(c \in d\backslash b \Leftrightarrow (c \in d \wedge c \notin b))$ | UI 2 |
| 43. | $(\forall c)(c \in d\backslash e \Leftrightarrow (c \in d \wedge c \notin e))$ | UI 42 |
| 44. | $g \in d\backslash e \Leftrightarrow (g \in d \wedge g \notin e)$ | UI 43 |
| 45. | $\neg(g \in d \wedge g \notin e)$ | NBE 41, 44 |
| 46. | $(\neg(g \in d)) \vee \neg(g \notin e)$ | DM 45 |
| 47. | $\neg(g \notin e)$ | ACP |
| 48. | $g \notin e \Leftrightarrow \neg(g \in e)$ | UI 39 |
| 49. | $\neg\neg(g \in e)$ | NBE 47, 48 |
| 50. | $g \in e$ | DN 49 |
| 51. | $(\neg(g \notin e)) \Rightarrow g \in e$ | CP 47-50 |
| 52. | $(\neg(g \in d)) \vee g \in e$ | D 46, 51 |
| 53. | $g \in d \Rightarrow g \in e$ | CE 52 |
| 54. | $(\forall c)(c \in d \Rightarrow c \in e)$ | UG 37-53 |
| 55. | $(\forall b)(d \subseteq b \Leftrightarrow (\forall c)(c \in d \Rightarrow c \in b))$ | UI 1 |
| 56. | $d \subseteq e \Leftrightarrow (\forall c)(c \in d \Rightarrow c \in e)$ | UI 55 |
| 57. | $d \subseteq e$ | BEl 54, 56 |
| 58. | $d\backslash e = \phi \Rightarrow d \subseteq e$ | CP 34-57 |
| 59. | $(d \subseteq e \Rightarrow d\backslash e = \phi) \wedge (d\backslash e = \phi \Rightarrow d \subseteq e)$ | C 33, 58 |
| 60. | $d \subseteq e \Leftrightarrow d\backslash e = \phi$ | BE 59 |
| 61. | $(\forall b)(d \subseteq b \Leftrightarrow d\backslash b = \phi)$ | UG 7-60 |
| 62. | $(\forall a)(\forall b)(a \subseteq b \Leftrightarrow a\backslash b = \phi)$ | UG 6-61 |

□

**Theorem 2.77** $(\forall a)(\forall b)(a \cap (b\backslash a) = \phi)$

***Proof***

| | | |
|---|---|---|
| 1. | $(\forall a)(\forall b)(\forall c)(c \in a \cap b \Leftrightarrow (c \in a \wedge c \in b))$ | Ax 2.15 |
| 2. | $(\forall a)(\forall b)(\forall c)(c \in a\backslash b \Leftrightarrow (c \in a \wedge c \notin b))$ | Ax 2.17 |
| 3. | $(\forall a)(\forall b)(a \notin b \Leftrightarrow \neg(a \in b))$ | Ax 2.12 |
| 4. | $(\forall a)((\forall b)(b \notin a) \Rightarrow a = \phi)$ | T 2.45 |
| 5. | flag $d$ | FSUG |
| 6. | flag $e$ | FSUG |
| 7. | flag $f$ | FSUG |
| 8. | $f \in d \cap (e\backslash d)$ | AIP |
| 9. | $(\forall b)(\forall c)(c \in d \cap b \Leftrightarrow (c \in d \wedge c \in b))$ | UI 1 |
| 10. | $(\forall c)(c \in d \cap (e\backslash d) \Leftrightarrow (c \in d \wedge c \in e\backslash d))$ | UI 9 |
| 11. | $f \in d \cap (e\backslash d) \Leftrightarrow (f \in d \wedge f \in e\backslash d)$ | UI 10 |
| 12. | $f \in d \wedge f \in e\backslash d$ | BEl 8, 11 |
| 13. | $f \in d$ | S 12 |
| 14. | $f \in e\backslash d$ | S 12 |
| 15. | $(\forall b)(\forall c)(c \in e\backslash b \Leftrightarrow (c \in e \wedge c \notin b))$ | UI 2 |
| 16. | $(\forall c)(c \in e\backslash d \Leftrightarrow (c \in e \wedge c \notin d))$ | UI 15 |
| 17. | $f \in e\backslash d \Leftrightarrow (f \in e \wedge f \notin d)$ | UI 16 |
| 18. | $f \in e \wedge f \notin d$ | BEl 14, 17 |
| 19. | $f \notin d$ | S 18 |
| 20. | $(\forall b)(f \notin b \Leftrightarrow \neg(f \in b))$ | UI 3 |
| 21. | $f \notin d \Leftrightarrow \neg(f \in d)$ | UI 20 |
| 22. | $\neg(f \in d)$ | BEl 19, 21 |
| 23. | $f \in d \wedge \neg(f \in d)$ | C 13, 22 |
| 24. | $\neg(f \in d \cap (e\backslash d))$ | IP 8-23 |
| 25. | $(\forall b)(f \notin b \Leftrightarrow \neg(f \in b))$ | UI 3 |
| 26. | $f \notin d \cap (e\backslash d) \Leftrightarrow \neg(f \in d \cap (e\backslash d))$ | UI 25 |
| 27. | $f \notin d \cap (e\backslash d)$ | BEl 24, 26 |
| 28. | $(\forall b)(b \notin d \cap (e\backslash d))$ | UG 7-27 |
| 29. | $(\forall b)(b \notin d \cap (e\backslash d)) \Rightarrow d \cap (e\backslash d) = \phi$ | UI 4 |
| 30. | $d \cap (e\backslash d) = \phi$ | MP 28, 29 |
| 31. | $(\forall b)(d \cap (b\backslash d) = \phi)$ | UG 6-30 |
| 32. | $(\forall a)(\forall b)(a \cap (b\backslash a) = \phi)$ | UG 5-31 |

□

**Theorem 2.78** $(\forall a)(\forall b)(a\backslash(a\backslash b) = a \cap b)$

***Proof***

| | | |
|---|---|---|
| 1. | $(\forall a)(\forall b)(\forall c)(c \in a\backslash b \Leftrightarrow (c \in a \wedge c \notin b))$ | Ax 2.17 |
| 2. | $(\forall a)(\forall b)(a \notin b \Leftrightarrow \neg(a \in b))$ | Ax 2.12 |
| 3. | $(\forall a)(\forall b)(\forall c)(c \in a \cap b \Leftrightarrow (c \in a \wedge c \in b))$ | Ax 2.15 |
| 4. | $(\forall a)(\forall b)((\forall c)(c \in a \Leftrightarrow c \in b) \Rightarrow a = b)$ | Ax 2.1 |
| 5. | flag $d$ | FSUG |
| 6. | flag $e$ | FSUG |
| 7. | flag $f$ | FSUG |
| 8. | $f \in d\backslash(d\backslash e)$ | ACP |
| 9. | $(\forall b)(\forall c)(c \in d\backslash b \Leftrightarrow (c \in d \wedge c \notin b))$ | UI 1 |
| 10. | $(\forall c)(c \in d\backslash(d\backslash e) \Leftrightarrow (c \in d \wedge c \notin d\backslash e))$ | UI 9 |
| 11. | $f \in d\backslash(d\backslash e) \Leftrightarrow (f \in d \wedge f \notin d\backslash e)$ | UI 10 |
| 12. | $f \in d \wedge f \notin d\backslash e$ | BEl 8, 11 |
| 13. | $f \in d$ | S 12 |
| 14. | $f \notin d\backslash e$ | S 12 |
| 15. | $(\forall b)(f \notin b \Leftrightarrow \neg(f \in b))$ | UI 2 |
| 16. | $f \notin d\backslash e \Leftrightarrow \neg(f \in d\backslash e)$ | UI 15 |
| 17. | $\neg(f \in d\backslash e)$ | BEl 14, 16 |
| 18. | $(\forall c)(c \in d\backslash e \Leftrightarrow (c \in d \wedge c \notin e))$ | UI 9 |
| 19. | $f \in d\backslash e \Leftrightarrow (f \in d \wedge f \notin e)$ | UI 18 |
| 20. | $\neg(f \in d \wedge f \notin e)$ | NBE 17, 19 |
| 21. | $(\neg(f \in d)) \vee \neg(f \notin e)$ | DM 20 |
| 22. | $f \in d \Rightarrow \neg(f \notin e)$ | CE 21 |
| 23. | $\neg(f \notin e)$ | MP 13, 22 |
| 24. | $f \notin e \Leftrightarrow \neg(f \in e)$ | UI 15 |
| 25. | $\neg\neg(f \in e)$ | NBE 23, 24 |
| 26. | $f \in e$ | DN 25 |
| 27. | $f \in d \wedge f \in e$ | C 13, 26 |
| 28. | $(\forall b)(\forall c)(c \in d \cap b \Leftrightarrow (c \in d \wedge c \in b))$ | UI 3 |
| 29. | $(\forall c)(c \in d \cap e \Leftrightarrow (c \in d \wedge c \in e))$ | UI 28 |
| 30. | $f \in d \cap e \Leftrightarrow (f \in d \wedge f \in e)$ | UI 29 |
| 31. | $f \in d \cap e$ | BEl 27, 30 |
| 32. | $f \in d\backslash(d\backslash e) \Rightarrow f \in d \cap e$ | CP 8-31 |
| 33. | $f \in d \cap e$ | ACP |
| 34. | $(\forall b)(\forall c)(c \in d \cap b \Leftrightarrow (c \in d \wedge c \in b))$ | UI 3 |
| 35. | $(\forall c)(c \in d \cap e \Leftrightarrow (c \in d \wedge c \in e))$ | UI 34 |
| 36. | $f \in d \cap e \Leftrightarrow (f \in d \wedge f \in e)$ | UI 35 |
| 37. | $f \in d \wedge f \in e$ | BEl 33, 36 |
| 38. | $f \in d$ | S 37 |
| 39. | $f \in e$ | S 37 |
| 40. | $\neg\neg(f \in e)$ | DN 39 |
| 41. | $(\forall b)(f \notin b \Leftrightarrow \neg(f \in b))$ | UI 2 |

| | | |
|---|---|---|
| 42. | $f \notin e \Leftrightarrow \neg(f \in e)$ | UI 41 |
| 43. | $\neg(f \notin e)$ | NBE 40, 42 |
| 44. | $(\neg(f \in d)) \vee \neg(f \notin e)$ | Ad 43 |
| 45. | $\neg(f \in d \wedge f \notin e)$ | DM 44 |
| 46. | $(\forall b)(\forall c)(c \in d\backslash b \Leftrightarrow (c \in d \wedge c \notin b))$ | UI 1 |
| 47. | $(\forall c)(c \in d\backslash e \Leftrightarrow (c \in d \wedge c \notin e))$ | UI 46 |
| 48. | $f \in d\backslash e \Leftrightarrow (f \in d \wedge f \notin e)$ | UI 47 |
| 49. | $\neg(f \in d\backslash e)$ | NBE 45, 48 |
| 50. | $f \notin d\backslash e \Leftrightarrow \neg(f \in d\backslash e)$ | UI 41 |
| 51. | $f \notin d\backslash e$ | BEI 49, 50 |
| 52. | $f \in d \wedge f \notin d\backslash e$ | C 38, 51 |
| 53. | $(\forall c)(c \in d\backslash(d\backslash e) \Leftrightarrow (c \in d \wedge c \notin d\backslash e))$ | UI 46 |
| 54. | $f \in d\backslash(d\backslash e) \Leftrightarrow (f \in d \wedge f \notin d\backslash e)$ | UI 53 |
| 55. | $f \in d\backslash(d\backslash e)$ | BEI 52, 54 |
| 56. | $f \in d \cap e \Rightarrow f \in d\backslash(d\backslash e)$ | CP 33-55 |
| 57. | $(f \in d\backslash(d\backslash e) \Rightarrow f \in d \cap e) \wedge$ $(f \in d \cap e \Rightarrow f \in d\backslash(d\backslash e))$ | C 32, 56 |
| 58. | $f \in d\backslash(d\backslash e) \Leftrightarrow f \in d \cap e$ | BE 57 |
| 59. | $(\forall c)(c \in d\backslash(d\backslash e) \Leftrightarrow c \in d \cap e)$ | UG 7-58 |
| 60. | $(\forall b)((\forall c)(c \in d\backslash(d\backslash e) \Leftrightarrow c \in b) \Rightarrow d\backslash(d\backslash e) = b)$ | UI 4 |
| 61. | $(\forall c)(c \in d\backslash(d\backslash e) \Leftrightarrow c \in d \cap e) \Rightarrow d\backslash(d\backslash e) = d \cap e$ | UI 60 |
| 62. | $d\backslash(d\backslash e) = d \cap e$ | MP 59, 61 |
| 63. | $(\forall b)(d\backslash(d\backslash b) = d \cap b)$ | UG 6-62 |
| 64. | $(\forall a)(\forall b)(a\backslash(a\backslash b) = a \cap b)$ | UG 5-63 |

□

## 2.8 Power Set

**Axiom 2.18 (Zermelo Axiom: Axiom of Power Set)** $(\forall a)(\exists b)(\forall c)(c \in b \Leftrightarrow c \subseteq a)$

**Theorem 2.79** $(\forall a)(\exists! b)(\forall c)(c \in b \Leftrightarrow c \subseteq a)$

***Proof***

| | | |
|---|---|---|
| 1. | $(\forall a)(\exists b)(\forall c)(c \in b \Leftrightarrow c \subseteq a)$ | Ax 2.18 |
| 2. | $(\forall a)(\forall b)((\forall c)(c \in a \Leftrightarrow c \in b) \Rightarrow a = b)$ | Ax 2.1 |
| 3. | flag $d$ | FSUG |
| 4. | $(\exists b)(\forall c)(c \in b \Leftrightarrow c \subseteq d)$ | UI 1 |
| 5. | $(\forall c)(c \in e \Leftrightarrow c \subseteq d)$ | EI 4, flag $e$ |

| | | |
|---|---|---|
| 6. | flag $f$ | FSUG |
| 7. | $(\forall c)(c \in f \Leftrightarrow c \subseteq d)$ | ACP |
| 8. | flag $g$ | FSUG |
| 9. | $g \in f \Leftrightarrow g \subseteq d$ | UI 7 |
| 10. | $g \in e \Leftrightarrow g \subseteq d$ | UI 5 |
| 11. | $g \in f \Leftrightarrow g \in e$ | BT 9, 10 |
| 12. | $(\forall c)(c \in f \Leftrightarrow c \in e)$ | UG 8-11 |
| 13. | $(\forall b)((\forall c)(c \in f \Leftrightarrow c \in b) \Rightarrow f = b)$ | UI 2 |
| 14. | $(\forall c)(c \in f \Leftrightarrow c \in e) \Rightarrow f = e$ | UI 13 |
| 15. | $f = e$ | MP 12, 14 |
| 16. | $(\forall c)(c \in f \Leftrightarrow c \subseteq d) \Rightarrow f = e$ | CP 7-15 |
| 17. | $(\forall h)((\forall c)(c \in h \Leftrightarrow c \subseteq d) \Rightarrow h = e)$ | UG 6-16 |
| 18. | $(\forall c)(c \in e \Leftrightarrow c \subseteq d) \wedge (\forall h)((\forall c)(c \in h \Leftrightarrow c \subseteq d) \Rightarrow h = e)$ | C 5, 17 |
| 19. | $(\exists b)((\forall c)(c \in b \Leftrightarrow c \subseteq d) \wedge$ $(\forall h)((\forall c)(c \in h \Leftrightarrow c \subseteq d) \Rightarrow h = b))$ | EG 18 |
| 20. | $(\exists! b)(\forall c)(c \in b \Leftrightarrow c \subseteq d)$ | UQ 19 |
| 21. | $(\forall a)(\exists! b)(\forall c)(c \in b \Leftrightarrow c \subseteq a)$ | UG 3-20 |

□

**Axiom 2.19 (Definitions of Operations, from Theorem 2.79, Which If Applied to the Argument $\alpha$ Give the Terms "P($\alpha$)" and "power set of $\alpha$")**

$(\forall a)(\forall b)(b \in \mathrm{P}(a) \Leftrightarrow b \subseteq a)$
$(\forall a)(\forall b)(b \in \text{power set of } a \Leftrightarrow b \subseteq a)$

**Theorem 2.80** $(\forall a)(a \in \mathrm{P}(a))$

***Proof***

| | | |
|---|---|---|
| 1. | $(\forall a)(a \subseteq a)$ | T 2.25 |
| 2. | $(\forall a)(\forall b)(b \in \mathrm{P}(a) \Leftrightarrow b \subseteq a)$ | Ax 2.19 |
| 3. | flag $c$ | FSUG |
| 4. | $c \subseteq c$ | UI 1 |
| 5. | $(\forall b)(b \in \mathrm{P}(c) \Leftrightarrow b \subseteq c)$ | UI 2 |
| 6. | $c \in \mathrm{P}(c) \Leftrightarrow c \subseteq c$ | UI 5 |
| 7. | $c \in \mathrm{P}(c)$ | BEl 4, 6 |
| 8. | $(\forall a)(a \in \mathrm{P}(a))$ | UG 3-7 |

□

**Theorem 2.81** $P(\phi) = \{\phi\}$

***Proof***

| | | |
|---|---|---|
| 1. | $(\forall a)(\forall b)(b \in P(a) \Leftrightarrow b \subseteq a)$ | Ax 2.19 |
| 2. | $(\forall a)(a \subseteq \phi \Rightarrow a = \phi)$ | T 2.43 |
| 3. | $(\forall a)(a \in \{a\})$ | T 2.6 |
| 4. | $(\forall a)(\forall b)(b \in \{a\} \Leftrightarrow a = b)$ | T 2.9 |
| 5. | $(\forall a)(a \subseteq a)$ | T 2.25 |
| 6. | $(\forall a)(\forall b)((\forall c)(c \in a \Leftrightarrow c \in b) \Rightarrow a = b)$ | Ax 2.1 |
| 7. | flag $d$ | FSUG |
| 8. | $(\forall b)(b \in P(\phi) \Leftrightarrow b \subseteq \phi)$ | UI 1 |
| 9. | $d \in P(\phi) \Leftrightarrow d \subseteq \phi$ | UI 8 |
| 10. | $d \subseteq \phi \Rightarrow d = \phi$ | UI 2 |
| 11. | $d \in P(\phi)$ | ACP |
| 12. | $d \subseteq \phi$ | BEI 9, 11 |
| 13. | $d = \phi$ | MP 10, 12 |
| 14. | $d \in \{d\}$ | UI 3 |
| 15. | $d \in \{\phi\}$ | IS 13, 14 |
| 16. | $d \in P(\phi) \Rightarrow d \in \{\phi\}$ | CP 11-15 |
| 17. | $d \in \{\phi\}$ | ACP |
| 18. | $(\forall b)(b \in \{\phi\} \Leftrightarrow \phi = b)$ | UI 4 |
| 19. | $d \in \{\phi\} \Leftrightarrow \phi = d$ | UI 18 |
| 20. | $\phi = d$ | BEI 17, 19 |
| 21. | $d \subseteq d$ | UI 5 |
| 22. | $d \subseteq \phi$ | IS 20, 21 |
| 23. | $d \in P(\phi)$ | BEI 9, 22 |
| 24. | $d \in \{\phi\} \Rightarrow d \in P(\phi)$ | CP 17-23 |
| 25. | $(d \in P(\phi) \Rightarrow d \in \{\phi\}) \wedge (d \in \{\phi\} \Rightarrow d \in P(\phi))$ | C 16, 24 |
| 26. | $d \in P(\phi) \Leftrightarrow d \in \{\phi\}$ | BE 25 |
| 27. | $(\forall c)(c \in P(\phi) \Leftrightarrow c \in \{\phi\})$ | UG 7-26 |
| 28. | $(\forall b)((\forall c)(c \in P(\phi) \Leftrightarrow c \in b) \Rightarrow P(\phi) = b)$ | UI 6 |
| 29. | $(\forall c)(c \in P(\phi) \Leftrightarrow c \in \{\phi\}) \Rightarrow P(\phi) = \{\phi\}$ | UI 28 |
| 30. | $P(\phi) = \{\phi\}$ | MP 27, 29 |

□

**Theorem 2.82** $(\forall a)(P(\{a\}) = \{\phi, \{a\}\})$

***Proof***

| | | |
|---|---|---|
| 1. | $(\forall a)(\phi \subseteq a)$ | T 2.42 |
| 2. | $(\forall a)(\forall b)(b \in \mathrm{P}(a) \Leftrightarrow b \subseteq a)$ | Ax 2.19 |
| 3. | $(\forall a)(a \neq \phi \Rightarrow (\exists b)(b \in a))$ | T 2.46 |
| 4. | $(\forall a)(\forall b)(a \subseteq b \Leftrightarrow (\forall c)(c \in a \Rightarrow c \in b))$ | Ax 2.9 |
| 5. | $(\forall a)(\forall b)(b \in \{a\} \Leftrightarrow a = b)$ | T 2.9 |
| 6. | $(\forall a)(\forall b)(a \in b \Leftrightarrow \{a\} \subseteq b)$ | T 2.32 |
| 7. | $(\forall a)(\forall b)(a = b \Leftrightarrow (a \subseteq b \wedge b \subseteq a))$ | T 2.24 |
| 8. | $(\forall a)(\forall b)(a = b \vee a \neq b)$ | T 2.15 |
| 9. | $(\forall a)(\forall b)(\forall c)(c \in \{a, b\} \Leftrightarrow (c = a \vee c = b))$ | Ax 2.3 |
| 10. | $(\forall a)(a \subseteq a)$ | T 2.25 |
| 11. | $(\forall a)(\forall b)((\forall c)(c \in a \Leftrightarrow c \in b) \Rightarrow a = b)$ | Ax 2.1 |
| 12. | flag $d$ | FSUG |
| 13. | $\phi \subseteq \{d\}$ | UI 1 |
| 14. | $(\forall b)(b \in \mathrm{P}(\{d\}) \Leftrightarrow b \subseteq \{d\})$ | UI 2 |
| 15. | $\phi \in \mathrm{P}(\{d\}) \Leftrightarrow \phi \subseteq \{d\}$ | UI 14 |
| 16. | flag $e$ | FSUG |
| 17. | $e \in \mathrm{P}(\{d\}) \wedge e \neq \phi$ | ACP |
| 18. | $e \neq \phi$ | S 17 |
| 19. | $e \neq \phi \Rightarrow (\exists b)(b \in e)$ | UI 3 |
| 20. | $(\exists b)(b \in e)$ | MP 18, 19 |
| 21. | $f \in e$ | EI 20, flag $f$ |
| 22. | $e \in \mathrm{P}(\{d\})$ | S 17 |
| 23. | $e \in \mathrm{P}(\{d\}) \Leftrightarrow e \subseteq \{d\}$ | UI 14 |
| 24. | $e \subseteq \{d\}$ | BEl 22, 23 |
| 25. | $(\forall b)(e \subseteq b \Leftrightarrow (\forall c)(c \in e \Rightarrow c \in b))$ | UI 4 |
| 26. | $e \subseteq \{d\} \Leftrightarrow (\forall c)(c \in e \Rightarrow c \in \{d\})$ | UI 25 |
| 27. | $(\forall c)(c \in e \Rightarrow c \in \{d\})$ | BEl 24, 26 |
| 28. | $f \in e \Rightarrow f \in \{d\}$ | UI 27 |
| 29. | $f \in \{d\}$ | MP 21, 28 |
| 30. | $(\forall b)(b \in \{d\} \Leftrightarrow d = b)$ | UI 5 |
| 31. | $f \in \{d\} \Leftrightarrow d = f$ | UI 30 |
| 32. | $d = f$ | BEl 29, 31 |
| 33. | $d \in e$ | IS 21, 32 |
| 34. | $(\forall b)(d \in b \Leftrightarrow \{d\} \subseteq b)$ | UI 6 |
| 35. | $d \in e \Leftrightarrow \{d\} \subseteq e$ | UI 34 |
| 36. | $\{d\} \subseteq e$ | BEl 33, 35 |
| 37. | $e \subseteq \{d\} \wedge \{d\} \subseteq e$ | C 24, 36 |
| 38. | $(\forall b)(e = b \Leftrightarrow (e \subseteq b \wedge b \subseteq e))$ | UI 7 |
| 39. | $e = \{d\} \Leftrightarrow (e \subseteq \{d\} \wedge \{d\} \subseteq e)$ | UI 38 |
| 40. | $e = \{d\}$ | BEl 37, 39 |

| | | |
|---|---|---|
| 41. | $(e \in \mathrm{P}(\{d\}) \wedge e \neq \phi) \Rightarrow e = \{d\}$ | CP 17-40 |
| 42. | $e \in \mathrm{P}(\{d\})$ | ACP |
| 43. | $(\forall b)(e = b \vee e \neq b)$ | UI 8 |
| 44. | $e = \phi \vee e \neq \phi$ | UI 43 |
| 45. | $e \neq \phi$ | ACP |
| 46. | $e \in \mathrm{P}(\{d\}) \wedge e \neq \phi$ | C 42, 45 |
| 47. | $e = \{d\}$ | MP 41, 46 |
| 48. | $e \neq \phi \Rightarrow e = \{d\}$ | CP 45-47 |
| 49. | $e = \phi \vee e = \{d\}$ | D 44, 48 |
| 50. | $e \in \mathrm{P}(\{d\}) \Rightarrow (e = \phi \vee e = \{d\})$ | CP 42-49 |
| 51. | $(\forall b)(\forall c)(c \in \{\phi, b\} \Leftrightarrow (c = \phi \vee c = b))$ | UI 9 |
| 52. | $(\forall c)(c \in \{\phi, \{d\}\} \Leftrightarrow (c = \phi \vee c = \{d\}))$ | UI 51 |
| 53. | $e \in \{\phi, \{d\}\} \Leftrightarrow (e = \phi \vee e = \{d\})$ | UI 52 |
| 54. | $(e \in \{\phi, \{d\}\} \Rightarrow (e = \phi \vee e = \{d\}))\wedge$ $((e = \phi \vee e = \{d\}) \Rightarrow e \in \{\phi, \{d\}\})$ | BE 53 |
| 55. | $(e = \phi \vee e = \{d\}) \Rightarrow e \in \{\phi, \{d\}\}$ | S 54 |
| 56. | $e \in \mathrm{P}(\{d\}) \Rightarrow e \in \{\phi, \{d\}\}$ | HS 50, 55 |
| 57. | $e \in \{\phi, \{d\}\}$ | ACP |
| 58. | $e = \phi \vee e = \{d\}$ | BEl 53, 57 |
| 59. | $\{d\} \subseteq \{d\}$ | UI 10 |
| 60. | $e = \phi$ | ACP |
| 61. | $e \subseteq \{d\}$ | IS 13, 60 |
| 62. | $e = \phi \Rightarrow e \subseteq \{d\}$ | CP 60-61 |
| 63. | $e = \{d\}$ | ACP |
| 64. | $e \subseteq \{d\}$ | IS 59, 63 |
| 65. | $e = \{d\} \Rightarrow e \subseteq \{d\}$ | CP 63-64 |
| 66. | $e \subseteq \{d\} \vee e \subseteq \{d\}$ | D 58, 62, 65 |
| 67. | $e \subseteq \{d\}$ | Du 66 |
| 68. | $e \in \mathrm{P}(\{d\}) \Leftrightarrow e \subseteq \{d\}$ | UI 14 |
| 69. | $e \in \mathrm{P}(\{d\})$ | BEl 67, 68 |
| 70. | $e \in \{\phi, \{d\}\} \Rightarrow e \in \mathrm{P}(\{d\})$ | CP 57-69 |
| 71. | $(e \in \mathrm{P}(\{d\}) \Rightarrow e \in \{\phi, \{d\}\})\wedge$ $(e \in \{\phi, \{d\}\} \Rightarrow e \in \mathrm{P}(\{d\}))$ | C 56, 70 |
| 72. | $e \in \mathrm{P}(\{d\}) \Leftrightarrow e \in \{\phi, \{d\}\}$ | BE 71 |
| 73. | $(\forall c)(c \in \mathrm{P}(\{d\}) \Leftrightarrow c \in \{\phi, \{d\}\})$ | UG 16-72 |
| 74. | $(\forall b)((\forall c)(c \in \mathrm{P}(\{d\}) \Leftrightarrow c \in b) \Rightarrow \mathrm{P}(\{d\}) = b)$ | UI 11 |
| 75. | $(\forall c)(c \in \mathrm{P}(\{d\}) \Leftrightarrow c \in \{\phi, \{d\}\}) \Rightarrow \mathrm{P}(\{d\}) = \{\phi, \{d\}\}$ | UI 74 |
| 76. | $\mathrm{P}(\{d\}) = \{\phi, \{d\}\}$ | MP 73, 75 |
| 77. | $(\forall a)(\mathrm{P}(\{a\}) = \{\phi, \{a\}\})$ | UG 12-76 |

□

## 2.9 Ordered Pair

**Theorem 2.83** $(\forall a)(\forall b)(\exists! c)(c = \{\{a\}, \{a, b\}\})$

***Proof***

| | | |
|---|---|---|
| 1. | flag $d$ | FSUG |
| 2. | flag $e$ | FSUG |
| 3. | $\{\{d\}, \{d, e\}\} = \{\{d\}, \{d, e\}\}$ | IR |
| 4. | $(\exists c)(c = \{\{d\}, \{d, e\}\})$ | EG 3 |
| 5. | flag $f$ | FSUG |
| 6. | flag $g$ | FSUG |
| 7. | $f = \{\{d\}, \{d, e\}\} \wedge g = \{\{d\}, \{d, e\}\}$ | ACP |
| 8. | $f = \{\{d\}, \{d, e\}\}$ | S 7 |
| 9. | $g = \{\{d\}, \{d, e\}\}$ | S 7 |
| 10. | $f = g$ | IT 8, 9 |
| 11. | $(f = \{\{d\}, \{d, e\}\} \wedge g = \{\{d\}, \{d, e\}\}) \Rightarrow f = g$ | CP 7-10 |
| 12. | $(\forall h)((f = \{\{d\}, \{d, e\}\} \wedge h = \{\{d\}, \{d, e\}\}) \Rightarrow f = h)$ | UG 6-11 |
| 13. | $(\forall i)(\forall h)((i = \{\{d\}, \{d, e\}\} \wedge h = \{\{d\}, \{d, e\}\}) \Rightarrow i = h)$ | UG 5-12 |
| 14. | $(\exists c)(c = \{\{d\}, \{d, e\}\}) \wedge$ $(\forall i)(\forall h)((i = \{\{d\}, \{d, e\}\} \wedge h = \{\{d\}, \{d, e\}\}) \Rightarrow i = h)$ | C 4, 13 |
| 15. | $(\exists! c)(c = \{\{d\}, \{d, e\}\})$ | UQ 14 |
| 16. | $(\forall b)(\exists! c)(c = \{\{d\}, \{d, b\}\})$ | UG 2-15 |
| 17. | $(\forall a)(\forall b)(\exists! c)(c = \{\{a\}, \{a, b\}\})$ | UG 1-16 |

□

**Axiom 2.20 (Definition of Operation, from Theorem 2.83, Which If Applied to the Arguments $\alpha$ and $\beta$ Gives the Term "$(\alpha, \beta)$")**
$(\forall a)(\forall b)((a, b) = \{\{a\}, \{a, b\}\})$

**Theorem 2.84** $(\forall a)(\forall b)(\forall c)(\forall d)((a, b) = (c, d) \Rightarrow a = c)$

***Proof***

| | | |
|---|---|---|
| 1. | $(\forall a)(\forall b)((a, b) = \{\{a\}, \{a, b\}\})$ | Ax 2.20 |
| 2. | $(\forall a)(\forall b)(a \in \{a, b\})$ | T 2.5 |
| 3. | $(\forall a)(\forall b)(\forall c)(c \in \{a, b\} \Leftrightarrow (c = a \vee c = b))$ | Ax 2.3 |
| 4. | $(\forall a)(\forall b)(\{a\} = \{b\} \Rightarrow a = b)$ | T 2.10 |
| 5. | $(\forall a)(\forall b)(\forall c)(\{a\} = \{b, c\} \Rightarrow a = b)$ | T 2.12 |
| 6. | flag $e$ | FSUG |
| 7. | flag $f$ | FSUG |
| 8. | flag $g$ | FSUG |
| 9. | flag $h$ | FSUG |
| 10. | $(e, f) = (g, h)$ | ACP |
| 11. | $(\forall b)((e, b) = \{\{e\}, \{e, b\}\})$ | UI 1 |

| | | |
|---|---|---|
| 12. | $(e, f) = \{\{e\}, \{e, f\}\}$ | UI 11 |
| 13. | $(\forall b)((g, b) = \{\{g\}, \{g, b\}\})$ | UI 1 |
| 14. | $(g, h) = \{\{g\}, \{g, h\}\}$ | UI 13 |
| 15. | $\{\{e\}, \{e, f\}\} = (g, h)$ | IS 10, 12 |
| 16. | $\{\{e\}, \{e, f\}\} = \{\{g\}, \{g, h\}\}$ | IS 14, 15 |
| 17. | $(\forall b)(\{e\} \in \{\{e\}, b\})$ | UI 2 |
| 18. | $\{e\} \in \{\{e\}, \{e, f\}\}$ | UI 17 |
| 19. | $\{e\} \in \{\{g\}, \{g, h\}\}$ | IS 16, 18 |
| 20. | $(\forall b)(\forall c)(c \in \{\{g\}, b\} \Leftrightarrow (c = \{g\} \vee c = b))$ | UI 3 |
| 21. | $(\forall c)(c \in \{\{g\}, \{g, h\}\} \Leftrightarrow (c = \{g\} \vee c = \{g, h\}))$ | UI 20 |
| 22. | $\{e\} \in \{\{g\}, \{g, h\}\}$ $\Leftrightarrow (\{e\} = \{g\} \vee \{e\} = \{g, h\})$ | UI 21 |
| 23. | $\{e\} = \{g\} \vee \{e\} = \{g, h\}$ | BEl 19, 22 |
| 24. | $(\forall b)(\{e\} = \{b\} \Rightarrow e = b)$ | UI 4 |
| 25. | $\{e\} = \{g\} \Rightarrow e = g$ | UI 24 |
| 26. | $(\forall b)(\forall c)(\{e\} = \{b, c\} \Rightarrow e = b)$ | UI 5 |
| 27. | $(\forall c)(\{e\} = \{g, c\} \Rightarrow e = g)$ | UI 26 |
| 28. | $\{e\} = \{g, h\} \Rightarrow e = g$ | UI 27 |
| 29. | $e = g \vee e = g$ | D 23, 25, 28 |
| 30. | $e = g$ | Du 29 |
| 31. | $(e, f) = (g, h) \Rightarrow e = g$ | CP 10-30 |
| 32. | $(\forall d)((e, f) = (g, d) \Rightarrow e = g)$ | UG 9-31 |
| 33. | $(\forall c)(\forall d)((e, f) = (c, d) \Rightarrow e = c)$ | UG 8-32 |
| 34. | $(\forall b)(\forall c)(\forall d)((e, b) = (c, d) \Rightarrow e = c)$ | UG 7-33 |
| 35. | $(\forall a)(\forall b)(\forall c)(\forall d)((a, b) = (c, d) \Rightarrow a = c)$ | UG 6-34 |

□

**Theorem 2.85** $(\forall a)(\forall b)(\forall c)(\forall d)((a, b) = (c, d) \Rightarrow b = d)$

***Proof***

| | | |
|---|---|---|
| 1. | $(\forall a)(\forall b)((a, b) = \{\{a\}, \{a, b\}\})$ | Ax 2.20 |
| 2. | $(\forall a)(\forall b)(a = b \vee a \neq b)$ | T 2.15 |
| 3. | $(\forall a)(\{a\} = \{a, a\})$ | Ax 2.4 |
| 4. | $(\forall a)(\forall b)(a \in \{b, a\})$ | T 2.7 |
| 5. | $(\forall a)(\forall b)(b \in \{a\} \Leftrightarrow a = b)$ | T 2.9 |
| 6. | $(\forall a)(\forall b)(\forall c)(\{a\} = \{b, c\} \Rightarrow a = c)$ | T 2.13 |
| 7. | $(\forall a)(\forall b)(a \neq b \Leftrightarrow \neg(a = b))$ | Ax 2.5 |
| 8. | $(\forall a)(\forall b)(\forall c)(c \in \{a, b\} \Leftrightarrow (c = a \vee c = b))$ | Ax 2.3 |
| 9. | $(\forall a)(\forall b)(\forall c)(\{a\} = \{b, c\} \Rightarrow b = c)$ | T 2.14 |
| 10. | $(\forall a)(\forall b)(\forall c)(\forall d)((a, b) = (c, d) \Rightarrow a = c)$ | T 2.84 |
| 11. | flag $e$ | FSUG |

| | | |
|---|---|---|
| 12. | flag $f$ | FSUG |
| 13. | flag $g$ | FSUG |
| 14. | flag $h$ | FSUG |
| 15. | $(e, f) = (g, h)$ | ACP |
| 16. | $(\forall b)((e, b) = \{\{e\}, \{e, b\}\})$ | UI 1 |
| 17. | $(e, f) = \{\{e\}, \{e, f\}\}$ | UI 16 |
| 18. | $(\forall b)((g, b) = \{\{g\}, \{g, b\}\})$ | UI 1 |
| 19. | $(g, h) = \{\{g\}, \{g, h\}\}$ | UI 18 |
| 20. | $\{\{e\}, \{e, f\}\} = (g, h)$ | IS 15, 17 |
| 21. | $\{\{e\}, \{e, f\}\} = \{\{g\}, \{g, h\}\}$ | IS 19, 20 |
| 22. | $(\forall b)(g = b \vee g \neq b)$ | UI 2 |
| 23. | $g = h \vee g \neq h$ | UI 22 |
| 24. | $g = h$ | ACP |
| 25. | $\{g, h\} = \{g, h\}$ | IR |
| 26. | $\{g, h\} = \{h, h\}$ | IS 24, 25 |
| 27. | $\{h\} = \{h, h\}$ | UI 3 |
| 28. | $\{g, h\} = \{h\}$ | IT 26, 27 |
| 29. | $\{\{g\}, \{g, h\}\} = \{\{g\}, \{g, h\}\}$ | IR |
| 30. | $\{\{g\}, \{g, h\}\} = \{\{h\}, \{g, h\}\}$ | IS 24, 29 |
| 31. | $\{\{g\}, \{g, h\}\} = \{\{h\}, \{h\}\}$ | IS 28, 30 |
| 32. | $\{\{h\}\} = \{\{h\}, \{h\}\}$ | UI 3 |
| 33. | $\{\{g\}, \{g, h\}\} = \{\{h\}\}$ | IT 31, 32 |
| 34. | $(\forall b)(\{e, f\} \in \{b, \{e, f\}\})$ | UI 4 |
| 35. | $\{e, f\} \in \{\{e\}, \{e, f\}\}$ | UI 34 |
| 36. | $\{e, f\} \in \{\{g\}, \{g, h\}\}$ | IS 21, 35 |
| 37. | $\{e, f\} \in \{\{h\}\}$ | IS 33, 36 |
| 38. | $(\forall b)(b \in \{\{h\}\} \Leftrightarrow \{h\} = b)$ | UI 5 |
| 39. | $\{e, f\} \in \{\{h\}\} \Leftrightarrow \{h\} = \{e, f\}$ | UI 38 |
| 40. | $\{h\} = \{e, f\}$ | BEl 37, 39 |
| 41. | $(\forall b)(\forall c)(\{h\} = \{b, c\} \Rightarrow h = c)$ | UI 6 |
| 42. | $(\forall c)(\{h\} = \{e, c\} \Rightarrow h = c)$ | UI 41 |
| 43. | $\{h\} = \{e, f\} \Rightarrow h = f$ | UI 42 |
| 44. | $h = f$ | MP 40, 43 |
| 45. | $g = h \Rightarrow h = f$ | CP 24-44 |
| 46. | $g \neq h$ | ACP |
| 47. | $(\forall b)(g \neq b \Leftrightarrow \neg(g = b))$ | UI 7 |
| 48. | $g \neq h \Leftrightarrow \neg(g = h)$ | UI 47 |
| 49. | $\neg(g = h)$ | BEl 46, 48 |
| 50. | $(\forall b)(\{g, h\} \in \{b, \{g, h\}\})$ | UI 4 |
| 51. | $\{g, h\} \in \{\{g\}, \{g, h\}\}$ | UI 50 |
| 52. | $\{g, h\} \in \{\{e\}, \{e, f\}\}$ | IS 21, 51 |
| 53. | $(\forall b)(\forall c)(c \in \{\{e\}, b\} \Leftrightarrow (c = \{e\} \vee c = b))$ | UI 8 |
| 54. | $(\forall c)(c \in \{\{e\}, \{e, f\}\} \Leftrightarrow$ $(c = \{e\} \vee c = \{e, f\}))$ | UI 53 |

| | | |
|---|---|---|
| 55. | $\{g, h\} \in \{\{e\}, \{e, f\}\} \Leftrightarrow$ $(\{g, h\} = \{e\} \vee \{g, h\} = \{e, f\})$ | UI 54 |
| 56. | $\{g, h\} = \{e\} \vee \{g, h\} = \{e, f\}$ | BEl 52, 55 |
| 57. | $\{g, h\} = \{e\}$ | AIP |
| 58. | $\{e\} = \{g, h\}$ | ISy 57 |
| 59. | $(\forall b)(\forall c)(\{e\} = \{b, c\} \Rightarrow b = c)$ | UI 9 |
| 60. | $(\forall c)(\{e\} = \{g, c\} \Rightarrow g = c)$ | UI 59 |
| 61. | $\{e\} = \{g, h\} \Rightarrow g = h$ | UI 60 |
| 62. | $g = h$ | MP 58, 61 |
| 63. | $g = h \wedge \neg(g = h)$ | C 49, 62 |
| 64. | $\neg(\{g, h\} = \{e\})$ | IP 57-63 |
| 65. | $\{g, h\} = \{e, f\}$ | DS 56, 64 |
| 66. | $(\forall b)(\forall c)(\forall d)((e, b) = (c, d) \Rightarrow e = c)$ | UI 10 |
| 67. | $(\forall c)(\forall d)((e, f) = (c, d) \Rightarrow e = c)$ | UI 66 |
| 68. | $(\forall d)((e, f) = (g, d) \Rightarrow e = g)$ | UI 67 |
| 69. | $(e, f) = (g, h) \Rightarrow e = g$ | UI 68 |
| 70. | $e = g$ | MP 15, 69 |
| 71. | $\{g, h\} = \{g, f\}$ | IS 65, 70 |
| 72. | $(\forall b)(h \in \{b, h\})$ | UI 4 |
| 73. | $h \in \{g, h\}$ | UI 72 |
| 74. | $h \in \{g, f\}$ | IS 71, 73 |
| 75. | $(\forall b)(\forall c)(c \in \{g, b\} \Leftrightarrow (c = g \vee c = b))$ | UI 8 |
| 76. | $(\forall c)(c \in \{g, f\} \Leftrightarrow (c = g \vee c = f))$ | UI 75 |
| 77. | $h \in \{g, f\} \Leftrightarrow (h = g \vee h = f)$ | UI 76 |
| 78. | $h = g \vee h = f$ | BEl 74, 77 |
| 79. | $h = g$ | ACP |
| 80. | $g = h$ | ISy 79 |
| 81. | $h = g \Rightarrow g = h$ | CP 79-80 |
| 82. | $g = h \vee h = f$ | D 78, 81 |
| 83. | $h = f$ | DS 49, 82 |
| 84. | $g \neq h \Rightarrow h = f$ | CP 46-83 |
| 85. | $h = f \vee h = f$ | D 23, 45, 84 |
| 86. | $h = f$ | Du 85 |
| 87. | $f = h$ | ISy 86 |
| 88. | $(e, f) = (g, h) \Rightarrow f = h$ | CP 15-87 |
| 89. | $(\forall d)((e, f) = (g, d) \Rightarrow f = d)$ | UG 14-88 |
| 90. | $(\forall c)(\forall d)((e, f) = (c, d) \Rightarrow f = d)$ | UG 13-89 |
| 91. | $(\forall b)(\forall c)(\forall d)((e, b) = (c, d) \Rightarrow b = d)$ | UG 12-90 |
| 92. | $(\forall a)(\forall b)(\forall c)(\forall d)((a, b) = (c, d) \Rightarrow b = d)$ | UG 11-91 |

□

**Theorem 2.86** $(\forall a)(\forall b)(\forall c)(\forall d)((a, b) = (c, d) \Rightarrow (a = c \wedge b = d))$

***Proof***

| | | |
|---|---|---|
| 1. | $(\forall a)(\forall b)(\forall c)(\forall d)((a, b) = (c, d) \Rightarrow a = c)$ | T 2.84 |
| 2. | $(\forall a)(\forall b)(\forall c)(\forall d)((a, b) = (c, d) \Rightarrow b = d)$ | T 2.85 |
| 3. | flag $e$ | FSUG |
| 4. | flag $f$ | FSUG |
| 5. | flag $g$ | FSUG |
| 6. | flag $h$ | FSUG |
| 7. | $(\forall b)(\forall c)(\forall d)((e, b) = (c, d) \Rightarrow e = c)$ | UI 1 |
| 8. | $(\forall c)(\forall d)((e, f) = (c, d) \Rightarrow e = c)$ | UI 7 |
| 9. | $(\forall d)((e, f) = (g, d) \Rightarrow e = g)$ | UI 8 |
| 10. | $(e, f) = (g, h) \Rightarrow e = g$ | UI 9 |
| 11. | $(\forall b)(\forall c)(\forall d)((e, b) = (c, d) \Rightarrow b = d)$ | UI 2 |
| 12. | $(\forall c)(\forall d)((e, f) = (c, d) \Rightarrow f = d)$ | UI 11 |
| 13. | $(\forall d)((e, f) = (g, d) \Rightarrow f = d)$ | UI 12 |
| 14. | $(e, f) = (g, h) \Rightarrow f = h$ | UI 13 |
| 15. | $(e, f) = (g, h)$ | ACP |
| 16. | $e = g$ | MP 10, 15 |
| 17. | $f = h$ | MP 14, 15 |
| 18. | $e = g \wedge f = h$ | C 16, 17 |
| 19. | $(e, f) = (g, h) \Rightarrow (e = g \wedge f = h)$ | CP 15-18 |
| 20. | $(\forall d)((e, f) = (g, d) \Rightarrow (e = g \wedge f = d))$ | UG 6-19 |
| 21. | $(\forall c)(\forall d)((e, f) = (c, d) \Rightarrow (e = c \wedge f = d))$ | UG 5-20 |
| 22. | $(\forall b)(\forall c)(\forall d)((e, b) = (c, d) \Rightarrow (e = c \wedge b = d))$ | UG 4-21 |
| 23. | $(\forall a)(\forall b)(\forall c)(\forall d)((a, b) = (c, d) \Rightarrow (a = c \wedge b = d))$ | UG 3-22 |

□

## 2.10 Cartesian Product

**Theorem 2.87** $(\forall a)(\forall b)(\forall c)(\forall d)((c \in a \wedge d \in b) \Rightarrow (c, d) \in \mathrm{P}(\mathrm{P}(a \cup b)))$

***Proof***

| | | |
|---|---|---|
| 1. | $(\forall a)(\forall b)(a \in b \Leftrightarrow \{a\} \subseteq b)$ | T 2.32 |
| 2. | $(\forall a)(\forall b)(a \subseteq a \cup b)$ | T 2.27 |
| 3. | $(\forall a)(\forall b)(\forall c)((a \subseteq b \wedge b \subseteq c) \Rightarrow a \subseteq c)$ | T 2.26 |
| 4. | $(\forall a)(\forall b)(b \in \mathrm{P}(a) \Leftrightarrow b \subseteq a)$ | Ax 2.19 |

| | | |
|---|---|---|
| 5. | $(\forall a)(\forall b)(\forall c)(\forall d)((a \in b \wedge c \in d) \Rightarrow \{a, c\} \subseteq b \cup d)$ | T 2.33 |
| 6. | $(\forall a)(\forall b)(\forall c)((a \in c \wedge b \in c) \Leftrightarrow \{a, b\} \subseteq c)$ | T 2.31 |
| 7. | $(\forall a)(\forall b)((a, b) = \{\{a\}, \{a, b\}\})$ | Ax 2.20 |
| 8. | flag $e$ | FSUG |
| 9. | flag $f$ | FSUG |
| 10. | flag $g$ | FSUG |
| 11. | flag $h$ | FSUG |
| 12. | $g \in e \wedge h \in f$ | ACP |
| 13. | $g \in e$ | S 12 |
| 14. | $(\forall b)(g \in b \Leftrightarrow \{g\} \subseteq b)$ | UI 1 |
| 15. | $g \in e \Leftrightarrow \{g\} \subseteq e$ | UI 14 |
| 16. | $\{g\} \subseteq e$ | BEl 13, 15 |
| 17. | $(\forall b)(e \subseteq e \cup b)$ | UI 2 |
| 18. | $e \subseteq e \cup f$ | UI 17 |
| 19. | $\{g\} \subseteq e \wedge e \subseteq e \cup f$ | C 16, 18 |
| 20. | $(\forall b)(\forall c)((\{g\} \subseteq b \wedge b \subseteq c) \Rightarrow \{g\} \subseteq c)$ | UI 3 |
| 21. | $(\forall c)((\{g\} \subseteq e \wedge e \subseteq c) \Rightarrow \{g\} \subseteq c)$ | UI 20 |
| 22. | $(\{g\} \subseteq e \wedge e \subseteq e \cup f) \Rightarrow \{g\} \subseteq e \cup f$ | UI 21 |
| 23. | $\{g\} \subseteq e \cup f$ | MP 19, 22 |
| 24. | $(\forall b)(b \in \mathrm{P}(e \cup f) \Leftrightarrow b \subseteq e \cup f)$ | UI 4 |
| 25. | $\{g\} \in \mathrm{P}(e \cup f) \Leftrightarrow \{g\} \subseteq e \cup f$ | UI 24 |
| 26. | $\{g\} \in \mathrm{P}(e \cup f)$ | BEl 23, 25 |
| 27. | $(\forall b)(\forall c)(\forall d)((g \in b \wedge c \in d) \Rightarrow \{g, c\} \subseteq b \cup d)$ | UI 5 |
| 28. | $(\forall c)(\forall d)((g \in e \wedge c \in d) \Rightarrow \{g, c\} \subseteq e \cup d)$ | UI 27 |
| 29. | $(\forall d)((g \in e \wedge h \in d) \Rightarrow \{g, h\} \subseteq e \cup d)$ | UI 28 |
| 30. | $(g \in e \wedge h \in f) \Rightarrow \{g, h\} \subseteq e \cup f$ | UI 29 |
| 31. | $\{g, h\} \subseteq e \cup f$ | MP 12, 30 |
| 32. | $\{g, h\} \in \mathrm{P}(e \cup f) \Leftrightarrow \{g, h\} \subseteq e \cup f$ | UI 24 |
| 33. | $\{g, h\} \in \mathrm{P}(e \cup f)$ | BEl 31, 32 |
| 34. | $\{g\} \in \mathrm{P}(e \cup f) \wedge \{g, h\} \in \mathrm{P}(e \cup f)$ | C 26, 33 |
| 35. | $(\forall b)(\forall c)((\{g\} \in c \wedge b \in c) \Leftrightarrow \{\{g\}, b\} \subseteq c)$ | UI 6 |
| 36. | $(\forall c)((\{g\} \in c \wedge \{g, h\} \in c) \Leftrightarrow \{\{g\}, \{g, h\}\} \subseteq c)$ | UI 35 |
| 37. | $(\{g\} \in \mathrm{P}(e \cup f) \wedge \{g, h\} \in \mathrm{P}(e \cup f)) \Leftrightarrow$ $\{\{g\}, \{g, h\}\} \subseteq \mathrm{P}(e \cup f)$ | UI 36 |
| 38. | $\{\{g\}, \{g, h\}\} \subseteq \mathrm{P}(e \cup f)$ | BEl 34, 37 |
| 39. | $(\forall b)((g, b) = \{\{g\}, \{g, b\}\})$ | UI 7 |
| 40. | $(g, h) = \{\{g\}, \{g, h\}\}$ | UI 39 |
| 41. | $(g, h) \subseteq \mathrm{P}(e \cup f)$ | IS 38, 40 |
| 42. | $(\forall b)(b \in \mathrm{P}(\mathrm{P}(e \cup f)) \Leftrightarrow b \subseteq \mathrm{P}(e \cup f))$ | UI 4 |
| 43. | $(g, h) \in \mathrm{P}(\mathrm{P}(e \cup f)) \Leftrightarrow (g, h) \subseteq \mathrm{P}(e \cup f)$ | UI 42 |
| 44. | $(g, h) \in \mathrm{P}(\mathrm{P}(e \cup f))$ | BEl 41, 43 |
| 45. | $(g \in e \wedge h \in f) \Rightarrow (g, h) \in \mathrm{P}(\mathrm{P}(e \cup f))$ | CP 12-44 |

| | | |
|---|---|---|
| 46. | $(\forall d)((g \in e \wedge d \in f) \Rightarrow (g, d) \in \mathrm{P}(\mathrm{P}(e \cup f)))$ | UG 11-45 |
| 47. | $(\forall c)(\forall d)((c \in e \wedge d \in f) \Rightarrow (c, d) \in \mathrm{P}(\mathrm{P}(e \cup f)))$ | UG 10-46 |
| 48. | $(\forall b)(\forall c)(\forall d)((c \in e \wedge d \in b) \Rightarrow (c, d) \in \mathrm{P}(\mathrm{P}(e \cup b)))$ | UG 9-47 |
| 49. | $(\forall a)(\forall b)(\forall c)(\forall d)((c \in a \wedge d \in b) \Rightarrow (c, d) \in \mathrm{P}(\mathrm{P}(a \cup b)))$ | UG 8-48 |

□

**Theorem 2.88** $(\forall a)(\forall b)(\exists c)(\forall d)(d \in c \Leftrightarrow (\exists e)(\exists f)(d = (e, f) \wedge (e \in a \wedge f \in b)))$

***Proof***

| | | |
|---|---|---|
| 1. | $(\forall a)(\forall b)(\forall g)(\exists c)(\forall d)(d \in c \Leftrightarrow$ $(d \in a \wedge (\exists e)(\exists f)(d = (e, f) \wedge (e \in g \wedge f \in b))))$ | AS 2.1 |
| 2. | $(\forall a)(\forall b)(\forall c)(\forall d)((c \in a \wedge d \in b) \Rightarrow (c, d) \in \mathrm{P}(\mathrm{P}(a \cup b)))$ | T 2.87 |
| 3. | flag $h$ | FSUG |
| 4. | flag $i$ | FSUG |
| 5. | $(\forall b)(\forall g)(\exists c)(\forall d)(d \in c \Leftrightarrow (d \in \mathrm{P}(\mathrm{P}(h \cup i)) \wedge$ $(\exists e)(\exists f)(d = (e, f) \wedge (e \in g \wedge f \in b))))$ | UI 1 |
| 6. | $(\forall g)(\exists c)(\forall d)(d \in c \Leftrightarrow (d \in \mathrm{P}(\mathrm{P}(h \cup i)) \wedge$ $(\exists e)(\exists f)(d = (e, f) \wedge (e \in g \wedge f \in i))))$ | UI 5 |
| 7. | $(\exists c)(\forall d)(d \in c \Leftrightarrow (d \in \mathrm{P}(\mathrm{P}(h \cup i)) \wedge$ $(\exists e)(\exists f)(d = (e, f) \wedge (e \in h \wedge f \in i))))$ | UI 6 |
| 8. | $(\forall d)(d \in j \Leftrightarrow (d \in \mathrm{P}(\mathrm{P}(h \cup i)) \wedge$ $(\exists e)(\exists f)(d = (e, f) \wedge (e \in h \wedge f \in i))))$ | EI 7, flag $j$ |
| 9. | flag $k$ | FSUG |
| 10. | $k \in j \Leftrightarrow (k \in \mathrm{P}(\mathrm{P}(h \cup i)) \wedge$ $(\exists e)(\exists f)(k = (e, f) \wedge (e \in h \wedge f \in i)))$ | UI 8 |
| 11. | $(k \in j \Rightarrow (k \in \mathrm{P}(\mathrm{P}(h \cup i)) \wedge$ $(\exists e)(\exists f)(k = (e, f) \wedge (e \in h \wedge f \in i)))) \wedge$ $((k \in \mathrm{P}(\mathrm{P}(h \cup i)) \wedge$ $(\exists e)(\exists f)(k = (e, f) \wedge (e \in h \wedge f \in i))) \Rightarrow k \in j)$ | BE 10 |
| 12. | $k \in j \Rightarrow (k \in \mathrm{P}(\mathrm{P}(h \cup i)) \wedge$ $(\exists e)(\exists f)(k = (e, f) \wedge (e \in h \wedge f \in i)))$ | S 11 |
| 13. | $(k \in \mathrm{P}(\mathrm{P}(h \cup i)) \wedge$ $(\exists e)(\exists f)(k = (e, f) \wedge (e \in h \wedge f \in i))) \Rightarrow k \in j$ | S 11 |
| 14. | $k \in j$ | ACP |
| 15. | $k \in \mathrm{P}(\mathrm{P}(h \cup i)) \wedge$ $(\exists e)(\exists f)(k = (e, f) \wedge (e \in h \wedge f \in i))$ | MP 12, 14 |
| 16. | $(\exists e)(\exists f)(k = (e, f) \wedge (e \in h \wedge f \in i))$ | S 15 |
| 17. | $k \in j \Rightarrow (\exists e)(\exists f)(k = (e, f) \wedge (e \in h \wedge f \in i))$ | CP 14-16 |
| 18. | $(\exists e)(\exists f)(k = (e, f) \wedge (e \in h \wedge f \in i))$ | ACP |
| 19. | $(\exists f)(k = (l, f) \wedge (l \in h \wedge f \in i))$ | EI 18, flag $l$ |

| | | |
|---|---|---|
| 20. | $k = (l, m) \wedge (l \in h \wedge m \in i)$ | EI 19, flag $m$ |
| 21. | $k = (l, m)$ | S 20 |
| 22. | $l \in h \wedge m \in i$ | S 20 |
| 23. | $(\forall b)(\forall c)(\forall d)$ $((c \in h \wedge d \in b) \Rightarrow (c, d) \in \mathrm{P}(\mathrm{P}(h \cup b)))$ | UI 2 |
| 24. | $(\forall c)(\forall d)((c \in h \wedge d \in i) \Rightarrow (c, d) \in \mathrm{P}(\mathrm{P}(h \cup i)))$ | UI 23 |
| 25. | $(\forall d)((l \in h \wedge d \in i) \Rightarrow (l, d) \in \mathrm{P}(\mathrm{P}(h \cup i)))$ | UI 24 |
| 26. | $(l \in h \wedge m \in i) \Rightarrow (l, m) \in \mathrm{P}(\mathrm{P}(h \cup i))$ | UI 25 |
| 27. | $(l, m) \in \mathrm{P}(\mathrm{P}(h \cup i))$ | MP 22, 26 |
| 28. | $k \in \mathrm{P}(\mathrm{P}(h \cup i))$ | IS 21, 27 |
| 29. | $k \in \mathrm{P}(\mathrm{P}(h \cup i)) \wedge$ $(\exists e)(\exists f)(k = (e, f) \wedge (e \in h \wedge f \in i))$ | C 18, 28 |
| 30. | $k \in j$ | MP 13, 29 |
| 31. | $(\exists e)(\exists f)(k = (e, f) \wedge (e \in h \wedge f \in i)) \Rightarrow k \in j$ | CP 18-30 |
| 32. | $(k \in j \Rightarrow (\exists e)(\exists f)(k = (e, f) \wedge (e \in h \wedge f \in i))) \wedge$ $((\exists e)(\exists f)(k = (e, f) \wedge (e \in h \wedge f \in i)) \Rightarrow k \in j)$ | C 17, 31 |
| 33. | $k \in j \Leftrightarrow (\exists e)(\exists f)(k = (e, f) \wedge (e \in h \wedge f \in i))$ | BE 32 |
| 34. | $(\forall d)(d \in j \Leftrightarrow (\exists e)(\exists f)(d = (e, f) \wedge (e \in h \wedge f \in i)))$ | UG 9-33 |
| 35. | $(\exists c)(\forall d)$ $(d \in c \Leftrightarrow (\exists e)(\exists f)(d = (e, f) \wedge (e \in h \wedge f \in i)))$ | EG 34 |
| 36. | $(\forall b)(\exists c)(\forall d)$ $(d \in c \Leftrightarrow (\exists e)(\exists f)(d = (e, f) \wedge (e \in h \wedge f \in b)))$ | UG 4-35 |
| 37. | $(\forall a)(\forall b)(\exists c)(\forall d)$ $(d \in c \Leftrightarrow (\exists e)(\exists f)(d = (e, f) \wedge (e \in a \wedge f \in b)))$ | UG 3-36 |

□

**Theorem 2.89** $(\forall a)(\forall b)(\exists! c)(\forall d)(d \in c \Leftrightarrow (\exists e)(\exists f)(d = (e, f) \wedge (e \in a \wedge f \in b)))$

***Proof***

| | | |
|---|---|---|
| 1. | $(\forall a)(\forall b)(\exists c)(\forall d)(d \in c \Leftrightarrow$ $(\exists e)(\exists f)(d = (e, f) \wedge (e \in a \wedge f \in b)))$ | T 2.88 |
| 2. | $(\forall a)(\forall b)((\forall c)(c \in a \Leftrightarrow c \in b) \Rightarrow a = b)$ | Ax 2.1 |
| 3. | flag $g$ | FSUG |
| 4. | flag $h$ | FSUG |
| 5. | $(\forall b)(\exists c)(\forall d)(d \in c \Leftrightarrow$ $(\exists e)(\exists f)(d = (e, f) \wedge (e \in g \wedge f \in b)))$ | UI 1 |
| 6. | $(\exists c)(\forall d)(d \in c \Leftrightarrow$ $(\exists e)(\exists f)(d = (e, f) \wedge (e \in g \wedge f \in h)))$ | UI 5 |
| 7. | $(\forall d)(d \in i \Leftrightarrow (\exists e)(\exists f)(d = (e, f) \wedge (e \in g \wedge f \in h)))$ | EI 6, flag $i$ |
| 8. | flag $j$ | FSUG |

| | | |
|---|---|---|
| 9. | $(\forall d)(d \in j \Leftrightarrow$ $(\exists e)(\exists f)(d = (e, f) \wedge (e \in g \wedge f \in h)))$ | ACP |
| 10. | flag $k$ | FSUG |
| 11. | $k \in j \Leftrightarrow (\exists e)(\exists f)(k = (e, f) \wedge (e \in g \wedge f \in h))$ | UI 9 |
| 12. | $k \in i \Leftrightarrow (\exists e)(\exists f)(k = (e, f) \wedge (e \in g \wedge f \in h))$ | UI 7 |
| 13. | $k \in j \Leftrightarrow k \in i$ | BT 11, 12 |
| 14. | $(\forall c)(c \in j \Leftrightarrow c \in i)$ | UG 10-13 |
| 15. | $(\forall b)((\forall c)(c \in j \Leftrightarrow c \in b) \Rightarrow j = b)$ | UI 2 |
| 16. | $(\forall c)(c \in j \Leftrightarrow c \in i) \Rightarrow j = i$ | UI 15 |
| 17. | $j = i$ | MP 14, 16 |
| 18. | $(\forall d)(d \in j \Leftrightarrow (\exists e)(\exists f)(d = (e, f) \wedge (e \in g \wedge f \in h)))$ $\Rightarrow j = i$ | CP 9-17 |
| 19. | $(\forall l)((\forall d)(d \in l \Leftrightarrow$ $(\exists e)(\exists f)(d = (e, f) \wedge (e \in g \wedge f \in h))) \Rightarrow l = i)$ | UG 8-18 |
| 20. | $(\forall d)(d \in i \Leftrightarrow (\exists e)(\exists f)(d = (e, f) \wedge (e \in g \wedge f \in h)))\wedge$ $(\forall l)((\forall d)(d \in l \Leftrightarrow$ $(\exists e)(\exists f)(d = (e, f) \wedge (e \in g \wedge f \in h))) \Rightarrow l = i)$ | C 7, 19 |
| 21. | $(\exists c)((\forall d)(d \in c \Leftrightarrow$ $(\exists e)(\exists f)(d = (e, f) \wedge (e \in g \wedge f \in h)))\wedge$ $(\forall l)((\forall d)(d \in l \Leftrightarrow$ $(\exists e)(\exists f)(d = (e, f) \wedge (e \in g \wedge f \in h))) \Rightarrow l = c))$ | EG 20 |
| 22. | $(\exists! c)(\forall d)(d \in c \Leftrightarrow$ $(\exists e)(\exists f)(d = (e, f) \wedge (e \in g \wedge f \in h)))$ | UQ 21 |
| 23. | $(\forall b)(\exists! c)(\forall d)(d \in c \Leftrightarrow$ $(\exists e)(\exists f)(d = (e, f) \wedge (e \in g \wedge f \in b)))$ | UG 4-22 |
| 24. | $(\forall a)(\forall b)(\exists! c)(\forall d)(d \in c \Leftrightarrow$ $(\exists e)(\exists f)(d = (e, f) \wedge (e \in a \wedge f \in b)))$ | UG 3-23 |

□

**Axiom 2.21 (Definitions of Operations, from Theorem 2.89, Which If Applied to the Arguments $\alpha$ and $\beta$ Give the Terms "$\alpha \times \beta$" and "Cartesian product of $\alpha$ and $\beta$")**

$(\forall a)(\forall b)(\forall c)(c \in a \times b \Leftrightarrow (\exists d)(\exists e)(c = (d, e) \wedge (d \in a \wedge e \in b)))$
$(\forall a)(\forall b)(\forall c)(c \in$ Cartesian product of $a$ and $b \Leftrightarrow (\exists d)(\exists e)(c = (d, e) \wedge (d \in a \wedge e \in b)))$

**Theorem 2.90** $(\forall a)(a \times \phi = \phi)$

***Proof***

| | | |
|---|---|---|
| 1. | $(\forall a)(\forall b)(a \neq b \Leftrightarrow \neg(a = b))$ | Ax 2.5 |
| 2. | $(\forall a)(a \neq \phi \Rightarrow (\exists b)(b \in a))$ | T 2.46 |
| 3. | $(\forall a)(\forall b)(\forall c)(c \in a \times b \Leftrightarrow$ | |
| | $(\exists d)(\exists e)(c = (d, e) \wedge (d \in a \wedge e \in b)))$ | Ax 2.21 |
| 4. | $(\forall a)(a \notin \phi)$ | Ax 2.14 |
| 5. | $(\forall a)(\forall b)(a \notin b \Leftrightarrow \neg(a \in b))$ | Ax 2.12 |
| 6. | flag $f$ | FSUG |
| 7. | $\neg(f \times \phi = \phi)$ | AIP |
| 8. | $(\forall b)(f \times \phi \neq b \Leftrightarrow \neg(f \times \phi = b))$ | UI 1 |
| 9. | $f \times \phi \neq \phi \Leftrightarrow \neg(f \times \phi = \phi)$ | UI 8 |
| 10. | $f \times \phi \neq \phi$ | BEl 7, 9 |
| 11. | $f \times \phi \neq \phi \Rightarrow (\exists b)(b \in f \times \phi)$ | UI 2 |
| 12. | $(\exists b)(b \in f \times \phi)$ | MP 10, 11 |
| 13. | $g \in f \times \phi$ | EI 12, flag $g$ |
| 14. | $(\forall b)(\forall c)(c \in f \times b \Leftrightarrow$ | |
| | $(\exists d)(\exists e)(c = (d, e) \wedge (d \in f \wedge e \in b)))$ | UI 3 |
| 15. | $(\forall c)(c \in f \times \phi \Leftrightarrow$ | |
| | $(\exists d)(\exists e)(c = (d, e) \wedge (d \in f \wedge e \in \phi)))$ | UI 14 |
| 16. | $g \in f \times \phi \Leftrightarrow (\exists d)(\exists e)(g = (d, e) \wedge (d \in f \wedge e \in \phi))$ | UI 15 |
| 17. | $(\exists d)(\exists e)(g = (d, e) \wedge (d \in f \wedge e \in \phi))$ | BEl 13, 16 |
| 18. | $(\exists e)(g = (h, e) \wedge (h \in f \wedge e \in \phi))$ | EI 17, flag $h$ |
| 19. | $g = (h, i) \wedge (h \in f \wedge i \in \phi)$ | EI 18, flag $i$ |
| 20. | $h \in f \wedge i \in \phi$ | S 19 |
| 21. | $i \in \phi$ | S 20 |
| 22. | $i \notin \phi$ | UI 4 |
| 23. | $(\forall b)(i \notin b \Leftrightarrow \neg(i \in b))$ | UI 5 |
| 24. | $i \notin \phi \Leftrightarrow \neg(i \in \phi)$ | UI 23 |
| 25. | $\neg(i \in \phi)$ | BEl 22, 24 |
| 26. | $i \in \phi \wedge \neg(i \in \phi)$ | C 21, 25 |
| 27. | $\neg\neg(f \times \phi = \phi)$ | IP 7-26 |
| 28. | $f \times \phi = \phi$ | DN 27 |
| 29. | $(\forall a)(a \times \phi = \phi)$ | UG 6-28 |

□

**Theorem 2.91** $(\forall a)(\phi \times a = \phi)$

***Proof***

| | | |
|---|---|---|
| 1. | $(\forall a)(\forall b)(a \neq b \Leftrightarrow \neg(a = b))$ | Ax 2.5 |
| 2. | $(\forall a)(a \neq \phi \Rightarrow (\exists b)(b \in a))$ | T 2.46 |
| 3. | $(\forall a)(\forall b)(\forall c)(c \in a \times b \Leftrightarrow$ | |
| | $(\exists d)(\exists e)(c = (d, e) \wedge (d \in a \wedge e \in b)))$ | Ax 2.21 |
| 4. | $(\forall a)(a \notin \phi)$ | Ax 2.14 |
| 5. | $(\forall a)(\forall b)(a \notin b \Leftrightarrow \neg(a \in b))$ | Ax 2.12 |
| 6. | flag $f$ | FSUG |
| 7. | $\neg(\phi \times f = \phi)$ | AIP |
| 8. | $(\forall b)(\phi \times f \neq b \Leftrightarrow \neg(\phi \times f = b))$ | UI 1 |
| 9. | $\phi \times f \neq \phi \Leftrightarrow \neg(\phi \times f = \phi)$ | UI 8 |
| 10. | $\phi \times f \neq \phi$ | BEl 7, 9 |
| 11. | $\phi \times f \neq \phi \Rightarrow (\exists b)(b \in \phi \times f)$ | UI 2 |
| 12. | $(\exists b)(b \in \phi \times f)$ | MP 10, 11 |
| 13. | $g \in \phi \times f$ | EI 12, flag $g$ |
| 14. | $(\forall b)(\forall c)(c \in \phi \times b \Leftrightarrow$ | |
| | $(\exists d)(\exists e)(c = (d, e) \wedge (d \in \phi \wedge e \in b)))$ | UI 3 |
| 15. | $(\forall c)(c \in \phi \times f \Leftrightarrow$ | |
| | $(\exists d)(\exists e)(c = (d, e) \wedge (d \in \phi \wedge e \in f)))$ | UI 14 |
| 16. | $g \in \phi \times f \Leftrightarrow (\exists d)(\exists e)(g = (d, e) \wedge (d \in \phi \wedge e \in f))$ | UI 15 |
| 17. | $(\exists d)(\exists e)(g = (d, e) \wedge (d \in \phi \wedge e \in f))$ | BEl 13, 16 |
| 18. | $(\exists e)(g = (h, e) \wedge (h \in \phi \wedge e \in f))$ | EI 17, flag $h$ |
| 19. | $g = (h, i) \wedge (h \in \phi \wedge i \in f)$ | EI 18, flag $i$ |
| 20. | $h \in \phi \wedge i \in f$ | S 19 |
| 21. | $h \in \phi$ | S 20 |
| 22. | $h \notin \phi$ | UI 4 |
| 23. | $(\forall b)(h \notin b \Leftrightarrow \neg(h \in b))$ | UI 5 |
| 24. | $h \notin \phi \Leftrightarrow \neg(h \in \phi)$ | UI 23 |
| 25. | $\neg(h \in \phi)$ | BEl 22, 24 |
| 26. | $h \in \phi \wedge \neg(h \in \phi)$ | C 21, 25 |
| 27. | $\neg\neg(\phi \times f = \phi)$ | IP 7-26 |
| 28. | $\phi \times f = \phi$ | DN 27 |
| 29. | $(\forall a)(\phi \times a = \phi)$ | UG 6-28 |

□

**Theorem 2.92** $(\forall a)(\forall b)(\forall c)(\forall d)((c \in a \wedge d \in b) \Leftrightarrow (c, d) \in a \times b)$

***Proof***

| | | |
|---|---|---|
| 1. | $(\forall a)(\forall b)(\forall c)(c \in a \times b \Leftrightarrow$ $(\exists e)(\exists f)(c = (e, f) \wedge (e \in a \wedge f \in b)))$ | Ax 2.21 |
| 2. | $(\forall a)(\forall b)(\forall c)(\forall d)((a, b) = (c, d) \Rightarrow (a = c \wedge b = d))$ | T 2.86 |
| 3. | flag $g$ | FSUG |
| 4. | flag $h$ | FSUG |
| 5. | flag $i$ | FSUG |
| 6. | flag $j$ | FSUG |
| 7. | $i \in g \wedge j \in h$ | ACP |
| 8. | $(i, j) = (i, j)$ | IR |
| 9. | $(i, j) = (i, j) \wedge (i \in g \wedge j \in h)$ | C 7, 8 |
| 10. | $(\exists f)((i, j) = (i, f) \wedge (i \in g \wedge f \in h))$ | EG 9 |
| 11. | $(\exists e)(\exists f)((i, j) = (e, f) \wedge (e \in g \wedge f \in h))$ | EG 10 |
| 12. | $(\forall b)(\forall c)(c \in g \times b \Leftrightarrow$ $(\exists e)(\exists f)(c = (e, f) \wedge (e \in g \wedge f \in b)))$ | UI 1 |
| 13. | $(\forall c)(c \in g \times h \Leftrightarrow$ $(\exists e)(\exists f)(c = (e, f) \wedge (e \in g \wedge f \in h)))$ | UI 12 |
| 14. | $(i, j) \in g \times h \Leftrightarrow$ $(\exists e)(\exists f)((i, j) = (e, f) \wedge (e \in g \wedge f \in h))$ | UI 13 |
| 15. | $(i, j) \in g \times h$ | BEI 11, 14 |
| 16. | $(i \in g \wedge j \in h) \Rightarrow (i, j) \in g \times h$ | CP 7-15 |
| 17. | $(i, j) \in g \times h$ | ACP |
| 18. | $(\forall b)(\forall c)(c \in g \times b \Leftrightarrow$ $(\exists e)(\exists f)(c = (e, f) \wedge (e \in g \wedge f \in b)))$ | UI 1 |
| 19. | $(\forall c)(c \in g \times h \Leftrightarrow$ $(\exists e)(\exists f)(c = (e, f) \wedge (e \in g \wedge f \in h)))$ | UI 18 |
| 20. | $(i, j) \in g \times h \Leftrightarrow$ $(\exists e)(\exists f)((i, j) = (e, f) \wedge (e \in g \wedge f \in h))$ | UI 19 |
| 21. | $(\exists e)(\exists f)((i, j) = (e, f) \wedge (e \in g \wedge f \in h))$ | BEI 17, 20 |
| 22. | $(\exists f)((i, j) = (k, f) \wedge (k \in g \wedge f \in h))$ | EI 21, flag $k$ |
| 23. | $(i, j) = (k, l) \wedge (k \in g \wedge l \in h)$ | EI 22, flag $l$ |
| 24. | $(i, j) = (k, l)$ | S 23 |
| 25. | $k \in g \wedge l \in h$ | S 23 |
| 26. | $(\forall b)(\forall c)(\forall d)((i, b) = (c, d) \Rightarrow (i = c \wedge b = d))$ | UI 2 |
| 27. | $(\forall c)(\forall d)((i, j) = (c, d) \Rightarrow (i = c \wedge j = d))$ | UI 26 |
| 28. | $(\forall d)((i, j) = (k, d) \Rightarrow (i = k \wedge j = d))$ | UI 27 |
| 29. | $(i, j) = (k, l) \Rightarrow (i = k \wedge j = l)$ | UI 28 |
| 30. | $i = k \wedge j = l$ | MP 24, 29 |
| 31. | $i = k$ | S 30 |
| 32. | $j = l$ | S 30 |
| 33. | $i \in g \wedge l \in h$ | IS 25, 31 |
| 34. | $i \in g \wedge j \in h$ | IS 32, 33 |
| 35. | $(i, j) \in g \times h \Rightarrow (i \in g \wedge j \in h)$ | CP 17-34 |

| | | |
|---|---|---|
| 36. | $((i \in g \wedge j \in h) \Rightarrow (i, j) \in g \times h) \wedge$ | |
| | $((i, j) \in g \times h \Rightarrow (i \in g \wedge j \in h))$ | C 16, 35 |
| 37. | $(i \in g \wedge j \in h) \Leftrightarrow (i, j) \in g \times h$ | BE 36 |
| 38. | $(\forall d)((i \in g \wedge d \in h) \Leftrightarrow (i, d) \in g \times h)$ | UG 6-37 |
| 39. | $(\forall c)(\forall d)((c \in g \wedge d \in h) \Leftrightarrow (c, d) \in g \times h)$ | UG 5-38 |
| 40. | $(\forall b)(\forall c)(\forall d)((c \in g \wedge d \in b) \Leftrightarrow (c, d) \in g \times b)$ | UG 4-39 |
| 41. | $(\forall a)(\forall b)(\forall c)(\forall d)((c \in a \wedge d \in b) \Leftrightarrow (c, d) \in a \times b)$ | UG 3-40 |

□

**Theorem 2.93** $(\forall a)(\forall b)((a = \phi \vee b = \phi) \Rightarrow a \times b = \phi)$

***Proof***

| | | |
|---|---|---|
| 1. | $(\forall a)(\phi \times a = \phi)$ | T 2.91 |
| 2. | $(\forall a)(a \times \phi = \phi)$ | T 2.90 |
| 3. | flag $c$ | FSUG |
| 4. | flag $d$ | FSUG |
| 5. | $c = \phi \vee d = \phi$ | ACP |
| 6. | $c = \phi$ | ACP |
| 7. | $c \times d = c \times d$ | IR |
| 8. | $c \times d = \phi \times d$ | IS 6, 7 |
| 9. | $\phi \times d = \phi$ | UI 1 |
| 10. | $c \times d = \phi$ | IT 8, 9 |
| 11. | $c = \phi \Rightarrow c \times d = \phi$ | CP 6-10 |
| 12. | $d = \phi$ | ACP |
| 13. | $c \times d = c \times d$ | IR |
| 14. | $c \times d = c \times \phi$ | IS 12, 13 |
| 15. | $c \times \phi = \phi$ | UI 2 |
| 16. | $c \times d = \phi$ | IT 14, 15 |
| 17. | $d = \phi \Rightarrow c \times d = \phi$ | CP 12-16 |
| 18. | $c \times d = \phi \vee c \times d = \phi$ | D 5, 11, 17 |
| 19. | $c \times d = \phi$ | Du 18 |
| 20. | $(c = \phi \vee d = \phi) \Rightarrow c \times d = \phi$ | CP 5-19 |
| 21. | $(\forall b)((c = \phi \vee b = \phi) \Rightarrow c \times b = \phi)$ | UG 4-20 |
| 22. | $(\forall a)(\forall b)((a = \phi \vee b = \phi) \Rightarrow a \times b = \phi)$ | UG 3-21 |

□

**Theorem 2.94** $(\forall a)(\forall b)((a = \phi \vee b = \phi) \Leftrightarrow a \times b = \phi)$

***Proof***

| | | |
|---|---|---|
| 1. | $(\forall a)(\forall b)(a \neq b \Leftrightarrow \neg(a = b))$ | Ax 2.5 |
| 2. | $(\forall a)(a \neq \phi \Rightarrow (\exists b)(b \in a))$ | T 2.46 |
| 3. | $(\forall a)(\forall b)(\forall c)(\forall d)((c \in a \wedge d \in b) \Leftrightarrow (c, d) \in a \times b)$ | T 2.92 |
| 4. | $(\forall a)((\exists b)(b \in a) \Rightarrow a \neq \phi)$ | T 2.48 |
| 5. | $(\forall a)(\forall b)((a = \phi \vee b = \phi) \Rightarrow a \times b = \phi)$ | T 2.93 |
| 6. | flag $e$ | FSUG |
| 7. | flag $f$ | FSUG |
| 8. | $e \times f = \phi$ | ACP |
| 9. | $\neg(e = \phi \vee f = \phi)$ | AIP |
| 10. | $(\neg(e = \phi)) \wedge \neg(f = \phi)$ | DM 9 |
| 11. | $\neg(e = \phi)$ | S 10 |
| 12. | $(\forall b)(e \neq b \Leftrightarrow \neg(e = b))$ | UI 1 |
| 13. | $e \neq \phi \Leftrightarrow \neg(e = \phi)$ | UI 12 |
| 14. | $e \neq \phi$ | BEl 11, 13 |
| 15. | $e \neq \phi \Rightarrow (\exists b)(b \in e)$ | UI 2 |
| 16. | $(\exists b)(b \in e)$ | MP 14, 15 |
| 17. | $g \in e$ | EI 16, flag $g$ |
| 18. | $\neg(f = \phi)$ | S 10 |
| 19. | $(\forall b)(f \neq b \Leftrightarrow \neg(f = b))$ | UI 1 |
| 20. | $f \neq \phi \Leftrightarrow \neg(f = \phi)$ | UI 19 |
| 21. | $f \neq \phi$ | BEl 18, 20 |
| 22. | $f \neq \phi \Rightarrow (\exists b)(b \in f)$ | UI 2 |
| 23. | $(\exists b)(b \in f)$ | MP 21, 22 |
| 24. | $h \in f$ | EI 23, flag $h$ |
| 25. | $g \in e \wedge h \in f$ | C 17, 24 |
| 26. | $(\forall b)(\forall c)(\forall d)((c \in e \wedge d \in b) \Leftrightarrow (c, d) \in e \times b)$ | UI 3 |
| 27. | $(\forall c)(\forall d)((c \in e \wedge d \in f) \Leftrightarrow (c, d) \in e \times f)$ | UI 26 |
| 28. | $(\forall d)((g \in e \wedge d \in f) \Leftrightarrow (g, d) \in e \times f)$ | UI 27 |
| 29. | $(g \in e \wedge h \in f) \Leftrightarrow (g, h) \in e \times f$ | UI 28 |
| 30. | $(g, h) \in e \times f$ | BEl 25, 29 |
| 31. | $(\exists b)(b \in e \times f)$ | EG 30 |
| 32. | $(\exists b)(b \in e \times f) \Rightarrow e \times f \neq \phi$ | UI 4 |
| 33. | $e \times f \neq \phi$ | MP 31, 32 |
| 34. | $(\forall b)(e \times f \neq b \Leftrightarrow \neg(e \times f = b))$ | UI 1 |
| 35. | $e \times f \neq \phi \Leftrightarrow \neg(e \times f = \phi)$ | UI 34 |
| 36. | $\neg(e \times f = \phi)$ | BEl 33, 35 |
| 37. | $e \times f = \phi \wedge \neg(e \times f = \phi)$ | C 8, 36 |
| 38. | $\neg\neg(e = \phi \vee f = \phi)$ | IP 9-37 |
| 39. | $e = \phi \vee f = \phi$ | DN 38 |

| | | |
|---|---|---|
| 40. | $e \times f = \phi \Rightarrow (e = \phi \vee f = \phi)$ | CP 8-39 |
| 41. | $(\forall b)((e = \phi \vee b = \phi) \Rightarrow e \times b = \phi)$ | UI 5 |
| 42. | $(e = \phi \vee f = \phi) \Rightarrow e \times f = \phi$ | UI 41 |
| 43. | $(e \times f = \phi \Rightarrow (e = \phi \vee f = \phi)) \wedge$ $((e = \phi \vee f = \phi) \Rightarrow e \times f = \phi)$ | C 40, 42 |
| 44. | $(e = \phi \vee f = \phi) \Leftrightarrow e \times f = \phi$ | BE 43 |
| 45. | $(\forall b)((e = \phi \vee b = \phi) \Leftrightarrow e \times b = \phi)$ | UG 7-44 |
| 46. | $(\forall a)(\forall b)((a = \phi \vee b = \phi) \Leftrightarrow a \times b = \phi)$ | UG 6-45 |

□

**Theorem 2.95** $(\forall a)(\forall b)(a \times b = \phi \Rightarrow b \times a = \phi)$

***Proof***

| | | |
|---|---|---|
| 1. | $(\forall a)(\forall b)((a = \phi \vee b = \phi) \Leftrightarrow a \times b = \phi)$ | T 2.94 |
| 2. | flag $c$ | FSUG |
| 3. | flag $d$ | FSUG |
| 4. | $c \times d = \phi$ | ACP |
| 5. | $(\forall b)((c = \phi \vee b = \phi) \Leftrightarrow c \times b = \phi)$ | UI 1 |
| 6. | $(c = \phi \vee d = \phi) \Leftrightarrow c \times d = \phi$ | UI 5 |
| 7. | $c = \phi \vee d = \phi$ | BEl 4, 6 |
| 8. | $d = \phi \vee c = \phi$ | Co 7 |
| 9. | $(\forall b)((d = \phi \vee b = \phi) \Leftrightarrow d \times b = \phi)$ | UI 1 |
| 10. | $(d = \phi \vee c = \phi) \Leftrightarrow d \times c = \phi$ | UI 9 |
| 11. | $d \times c = \phi$ | BEl 8, 10 |
| 12. | $c \times d = \phi \Rightarrow d \times c = \phi$ | CP 4-11 |
| 13. | $(\forall b)(c \times b = \phi \Rightarrow b \times c = \phi)$ | UG 3-12 |
| 14. | $(\forall a)(\forall b)(a \times b = \phi \Rightarrow b \times a = \phi)$ | UG 2-13 |

□

**Theorem 2.96** $(\forall a)(\forall b)(a \times b = b \times a \Leftrightarrow (a = b \vee (a = \phi \vee b = \phi)))$

***Proof***

| | | |
|---|---|---|
| 1. | $(\forall a)(\forall b)((a = \phi \vee b = \phi) \Leftrightarrow a \times b = \phi)$ | T 2.94 |
| 2. | $(\forall a)(\forall b)(a \times b = \phi \Rightarrow b \times a = \phi)$ | T 2.95 |
| 3. | $(\forall a)(\forall b)(a = b \Rightarrow (\forall c)(c \in a \Leftrightarrow c \in b))$ | T 2.1 |
| 4. | $(\forall a)(\forall b)(a \neq b \Leftrightarrow \neg(a = b))$ | Ax 2.5 |
| 5. | $(\forall a)(a \neq \phi \Rightarrow (\exists b)(b \in a))$ | T 2.46 |
| 6. | $(\forall a)(\forall b)(\forall c)(\forall d)((c \in a \wedge d \in b) \Leftrightarrow (c, d) \in a \times b)$ | T 2.92 |
| 7. | $(\forall a)(\forall b)((\forall c)(c \in a \Leftrightarrow c \in b) \Rightarrow a = b)$ | Ax 2.1 |
| 8. | flag $e$ | FSUG |

| | | |
|---|---|---|
| 9. | flag $f$ | FSUG |
| 10. | $(\forall b)((e = \phi \vee b = \phi) \Leftrightarrow e \times b = \phi)$ | UI 1 |
| 11. | $(e = \phi \vee f = \phi) \Leftrightarrow e \times f = \phi$ | UI 10 |
| 12. | $\quad e = f \vee (e = \phi \vee f = \phi)$ | ACP |
| 13. | $\quad\quad e = f$ | ACP |
| 14. | $\quad\quad e \times e = e \times e$ | IR |
| 15. | $\quad\quad e \times f = f \times e$ | IS 13, 14 |
| 16. | $\quad e = f \Rightarrow e \times f = f \times e$ | CP 13-15 |
| 17. | $\quad\quad e = \phi \vee f = \phi$ | ACP |
| 18. | $\quad\quad e \times f = \phi$ | BEI 11, 17 |
| 19. | $\quad\quad (\forall b)(e \times b = \phi \Rightarrow b \times e = \phi)$ | UI 2 |
| 20. | $\quad\quad e \times f = \phi \Rightarrow f \times e = \phi$ | UI 19 |
| 21. | $\quad\quad f \times e = \phi$ | MP 18, 20 |
| 22. | $\quad\quad e \times f = f \times e$ | IT 18, 21 |
| 23. | $\quad (e = \phi \vee f = \phi) \Rightarrow e \times f = f \times e$ | CP 17-22 |
| 24. | $\quad e \times f = f \times e \vee e \times f = f \times e$ | D 12, 16, 23 |
| 25. | $\quad e \times f = f \times e$ | Du 24 |
| 26. | $(e = f \vee (e = \phi \vee f = \phi)) \Rightarrow e \times f = f \times e$ | CP 12-25 |
| 27. | $\quad e \times f = f \times e$ | ACP |
| 28. | $\quad e \times f = \phi \vee \neg(e \times f = \phi)$ | EM |
| 29. | $\quad\quad e \times f = \phi$ | ACP |
| 30. | $\quad\quad e = \phi \vee f = \phi$ | BEI 11, 29 |
| 31. | $\quad e \times f = \phi \Rightarrow (e = \phi \vee f = \phi)$ | CP 29-30 |
| 32. | $\quad (\forall b)(e \times f = b \Rightarrow (\forall c)(c \in e \times f \Leftrightarrow c \in b))$ | UI 3 |
| 33. | $\quad e \times f = f \times e \Rightarrow (\forall c)(c \in e \times f \Leftrightarrow c \in f \times e)$ | UI 32 |
| 34. | $\quad (\forall c)(c \in e \times f \Leftrightarrow c \in f \times e)$ | MP 27, 33 |
| 35. | $\quad\quad \neg(e \times f = \phi)$ | ACP |
| 36. | $\quad\quad \neg(e = \phi \vee f = \phi)$ | NBE 11, 35 |
| 37. | $\quad\quad (\neg(e = \phi)) \wedge \neg(f = \phi)$ | DM 36 |
| 38. | $\quad\quad \neg(e = \phi)$ | S 37 |
| 39. | $\quad\quad \neg(f = \phi)$ | S 37 |
| 40. | $\quad\quad (\forall b)(f \neq b \Leftrightarrow \neg(f = b))$ | UI 4 |
| 41. | $\quad\quad f \neq \phi \Leftrightarrow \neg(f = \phi)$ | UI 40 |
| 42. | $\quad\quad f \neq \phi$ | BEI 39, 41 |
| 43. | $\quad\quad f \neq \phi \Rightarrow (\exists b)(b \in f)$ | UI 5 |
| 44. | $\quad\quad (\exists b)(b \in f)$ | MP 42, 43 |
| 45. | $\quad\quad g \in f$ | EI 44, flag $g$ |
| 46. | $\quad\quad (\forall b)(e \neq b \Leftrightarrow \neg(e = b))$ | UI 4 |
| 47. | $\quad\quad e \neq \phi \Leftrightarrow \neg(e = \phi)$ | UI 46 |
| 48. | $\quad\quad e \neq \phi$ | BEI 38, 47 |
| 49. | $\quad\quad e \neq \phi \Rightarrow (\exists b)(b \in e)$ | UI 5 |
| 50. | $\quad\quad (\exists b)(b \in e)$ | MP 48, 49 |

| | | |
|---|---|---|
| 51. | $h \in e$ | EI 50, flag $h$ |
| 52. | flag $i$ | FSUG |
| 53. | $i \in e$ | ACP |
| 54. | $i \in e \wedge g \in f$ | C 45, 53 |
| 55. | $(\forall b)(\forall c)(\forall d)$ | |
| | $((c \in e \wedge d \in b) \Leftrightarrow (c, d) \in e \times b)$ | UI 6 |
| 56. | $(\forall c)(\forall d)((c \in e \wedge d \in f) \Leftrightarrow (c, d) \in e \times f)$ | UI 55 |
| 57. | $(\forall d)((i \in e \wedge d \in f) \Leftrightarrow (i, d) \in e \times f)$ | UI 56 |
| 58. | $(i \in e \wedge g \in f) \Leftrightarrow (i, g) \in e \times f$ | UI 57 |
| 59. | $(i, g) \in e \times f$ | BEl 54, 58 |
| 60. | $(i, g) \in e \times f \Leftrightarrow (i, g) \in f \times e$ | UI 34 |
| 61. | $(i, g) \in f \times e$ | BEl 59, 60 |
| 62. | $(\forall b)(\forall c)(\forall d)$ | |
| | $((c \in f \wedge d \in b) \Leftrightarrow (c, d) \in f \times b)$ | UI 6 |
| 63. | $(\forall c)(\forall d)((c \in f \wedge d \in e) \Leftrightarrow (c, d) \in f \times e)$ | UI 62 |
| 64. | $(\forall d)((i \in f \wedge d \in e) \Leftrightarrow (i, d) \in f \times e)$ | UI 63 |
| 65. | $(i \in f \wedge g \in e) \Leftrightarrow (i, g) \in f \times e$ | UI 64 |
| 66. | $i \in f \wedge g \in e$ | BEl 61, 65 |
| 67. | $i \in f$ | S 66 |
| 68. | $i \in e \Rightarrow i \in f$ | CP 53-67 |
| 69. | $i \in f$ | ACP |
| 70. | $h \in e \wedge i \in f$ | C 51, 69 |
| 71. | $(\forall b)(\forall c)(\forall d)$ | |
| | $((c \in e \wedge d \in b) \Leftrightarrow (c, d) \in e \times b)$ | UI 6 |
| 72. | $(\forall c)(\forall d)((c \in e \wedge d \in f) \Leftrightarrow (c, d) \in e \times f)$ | UI 71 |
| 73. | $(\forall d)((h \in e \wedge d \in f) \Leftrightarrow (h, d) \in e \times f)$ | UI 72 |
| 74. | $(h \in e \wedge i \in f) \Leftrightarrow (h, i) \in e \times f$ | UI 73 |
| 75. | $(h, i) \in e \times f$ | BEl 70, 74 |
| 76. | $(h, i) \in e \times f \Leftrightarrow (h, i) \in f \times e$ | UI 34 |
| 77. | $(h, i) \in f \times e$ | BEl 75, 76 |
| 78. | $(\forall b)(\forall c)(\forall d)$ | |
| | $((c \in f \wedge d \in b) \Leftrightarrow (c, d) \in f \times b)$ | UI 6 |
| 79. | $(\forall c)(\forall d)((c \in f \wedge d \in e) \Leftrightarrow (c, d) \in f \times e)$ | UI 78 |
| 80. | $(\forall d)((h \in f \wedge d \in e) \Leftrightarrow (h, d) \in f \times e)$ | UI 79 |
| 81. | $(h \in f \wedge i \in e) \Leftrightarrow (h, i) \in f \times e$ | UI 80 |
| 82. | $h \in f \wedge i \in e$ | BEl 77, 81 |
| 83. | $i \in e$ | S 82 |
| 84. | $i \in f \Rightarrow i \in e$ | CP 69-83 |
| 85. | $(i \in e \Rightarrow i \in f) \wedge (i \in f \Rightarrow i \in e)$ | C 68, 84 |
| 86. | $i \in e \Leftrightarrow i \in f$ | BE 85 |
| 87. | $(\forall c)(c \in e \Leftrightarrow c \in f)$ | UG 52-86 |
| 88. | $(\forall b)((\forall c)(c \in e \Leftrightarrow c \in b) \Rightarrow e = b)$ | UI 7 |

| | | |
|---|---|---|
| 89. | $(\forall c)(c \in e \Leftrightarrow c \in f) \Rightarrow e = f$ | UI 88 |
| 90. | $e = f$ | MP 87, 89 |
| 91. | $(\neg(e \times f = \phi)) \Rightarrow e = f$ | CP 35-90 |
| 92. | $e = f \vee (e = \phi \vee f = \phi)$ | D 28, 31, 91 |
| 93. | $e \times f = f \times e \Rightarrow (e = f \vee (e = \phi \vee f = \phi))$ | CP 27-92 |
| 94. | $(e \times f = f \times e \Rightarrow (e = f \vee (e = \phi \vee f = \phi)))\wedge$ $((e = f \vee (e = \phi \vee f = \phi)) \Rightarrow e \times f = f \times e)$ | C 26, 93 |
| 95. | $e \times f = f \times e \Leftrightarrow (e = f \vee (e = \phi \vee f = \phi))$ | BE 94 |
| 96. | $(\forall b)(e \times b = b \times e \Leftrightarrow (e = b \vee (e = \phi \vee b = \phi)))$ | UG 9-95 |
| 97. | $(\forall a)(\forall b)(a \times b = b \times a \Leftrightarrow (a = b \vee (a = \phi \vee b = \phi)))$ | UG 8-96 |

□

## 2.11 Relation and Function

**Axiom 2.22 (Definition of Predicate Which If Applied to the Argument $\alpha$ Gives the Atomic Formula "$\alpha$ is a relation")**
$(\forall a)(a \text{ is a relation} \Leftrightarrow (\forall b)(b \in a \Rightarrow (\exists c)(\exists d)(b = (c, d))))$

**Theorem 2.97** $(\forall a)(\forall b)(a \times b \text{ is a relation})$

***Proof***

| | | |
|---|---|---|
| 1. | $(\forall a)(\forall b)(\forall c)(c \in a \times b \Leftrightarrow$ $(\exists d)(\exists e)(c = (d, e) \wedge (d \in a \wedge e \in b)))$ | Ax 2.21 |
| 2. | $(\forall a)(a \text{ is a relation} \Leftrightarrow (\forall b)(b \in a \Rightarrow (\exists c)(\exists d)(b = (c, d))))$ | Ax 2.22 |
| 3. | flag $f$ | FSUG |
| 4. | flag $g$ | FSUG |
| 5. | flag $h$ | FSUG |
| 6. | $h \in f \times g$ | ACP |
| 7. | $(\forall b)(\forall c)(c \in f \times b \Leftrightarrow$ $(\exists d)(\exists e)(c = (d, e) \wedge (d \in f \wedge e \in b)))$ | UI 1 |
| 8. | $(\forall c)(c \in f \times g \Leftrightarrow$ $(\exists d)(\exists e)(c = (d, e) \wedge (d \in f \wedge e \in g)))$ | UI 7 |
| 9. | $h \in f \times g \Leftrightarrow$ $(\exists d)(\exists e)(h = (d, e) \wedge (d \in f \wedge e \in g))$ | UI 8 |
| 10. | $(\exists d)(\exists e)(h = (d, e) \wedge (d \in f \wedge e \in g))$ | BEI 6, 9 |
| 11. | $(\exists e)(h = (i, e) \wedge (i \in f \wedge e \in g))$ | EI 10, flag $i$ |
| 12. | $h = (i, j) \wedge (i \in f \wedge j \in g)$ | EI 11, flag $j$ |
| 13. | $h = (i, j)$ | S 12 |
| 14. | $(\exists d)(h = (i, d))$ | EG 13 |

| | | |
|---|---|---|
| 15. | $(\exists c)(\exists d)(h = (c, d))$ | EG 14 |
| 16. | $h \in f \times g \Rightarrow (\exists c)(\exists d)(h = (c, d))$ | CP 6-15 |
| 17. | $(\forall b)(b \in f \times g \Rightarrow (\exists c)(\exists d)(b = (c, d)))$ | UG 5-16 |
| 18. | $f \times g$ is a relation $\Leftrightarrow$ | |
| | $(\forall b)(b \in f \times g \Rightarrow (\exists c)(\exists d)(b = (c, d)))$ | UI 2 |
| 19. | $f \times g$ is a relation | BEI 17, 18 |
| 20. | $(\forall b)(f \times b$ is a relation$)$ | UG 4-19 |
| 21. | $(\forall a)(\forall b)(a \times b$ is a relation$)$ | UG 3-20 |

□

**Theorem 2.98** $(\forall a)(\forall b)(\forall c)(\{b, c\} \in a \Rightarrow (b \in \bigcup a \wedge c \in \bigcup a))$

***Proof***

| | | |
|---|---|---|
| 1. | $(\forall a)(\forall b)(a \in \{a, b\})$ | T 2.5 |
| 2. | $(\forall a)(\forall b)(b \in \bigcup a \Leftrightarrow (\exists c)(b \in c \wedge c \in a))$ | Ax 2.7 |
| 3. | $(\forall a)(\forall b)(a \in \{b, a\})$ | T 2.7 |
| 4. | flag $d$ | FSUG |
| 5. | flag $e$ | FSUG |
| 6. | flag $f$ | FSUG |
| 7. | $\{e, f\} \in d$ | ACP |
| 8. | $(\forall b)(e \in \{e, b\})$ | UI 1 |
| 9. | $e \in \{e, f\}$ | UI 8 |
| 10. | $e \in \{e, f\} \wedge \{e, f\} \in d$ | C 7, 9 |
| 11. | $(\exists c)(e \in c \wedge c \in d)$ | EG 10 |
| 12. | $(\forall b)(b \in \bigcup d \Leftrightarrow (\exists c)(b \in c \wedge c \in d))$ | UI 2 |
| 13. | $e \in \bigcup d \Leftrightarrow (\exists c)(e \in c \wedge c \in d)$ | UI 12 |
| 14. | $e \in \bigcup d$ | BEI 11, 13 |
| 15. | $(\forall b)(f \in \{b, f\})$ | UI 3 |
| 16. | $f \in \{e, f\}$ | UI 15 |
| 17. | $f \in \{e, f\} \wedge \{e, f\} \in d$ | C 7, 16 |
| 18. | $(\exists c)(f \in c \wedge c \in d)$ | EG 17 |
| 19. | $f \in \bigcup d \Leftrightarrow (\exists c)(f \in c \wedge c \in d)$ | UI 12 |
| 20. | $f \in \bigcup d$ | BEI 18, 19 |
| 21. | $e \in \bigcup d \wedge f \in \bigcup d$ | C 14, 20 |
| 22. | $\{e, f\} \in d \Rightarrow (e \in \bigcup d \wedge f \in \bigcup d)$ | CP 7-21 |
| 23. | $(\forall c)(\{e, c\} \in d \Rightarrow (e \in \bigcup d \wedge c \in \bigcup d))$ | UG 6-22 |
| 24. | $(\forall b)(\forall c)(\{b, c\} \in d \Rightarrow (b \in \bigcup d \wedge c \in \bigcup d))$ | UG 5-23 |
| 25. | $(\forall a)(\forall b)(\forall c)(\{b, c\} \in a \Rightarrow (b \in \bigcup a \wedge c \in \bigcup a))$ | UG 4-24 |

□

**Theorem 2.99** $(\forall a)(\forall b)(\forall c)((b,c) \in a \Rightarrow (b \in \cup\cup a \wedge c \in \cup\cup a))$

***Proof***

| | | |
|---|---|---|
| 1. | $(\forall a)(\forall b)((a,b) = \{\{a\},\{a,b\}\})$ | Ax 2.20 |
| 2. | $(\forall a)(\forall b)(\forall c)(\{b,c\} \in a \Rightarrow (b \in \cup a \wedge c \in \cup a))$ | T 2.98 |
| 3. | flag $d$ | FSUG |
| 4. | flag $e$ | FSUG |
| 5. | flag $f$ | FSUG |
| 6. | $(e,f) \in d$ | ACP |
| 7. | $(\forall b)((e,b) = \{\{e\},\{e,b\}\})$ | UI 1 |
| 8. | $(e,f) = \{\{e\},\{e,f\}\}$ | UI 7 |
| 9. | $\{\{e\},\{e,f\}\} \in d$ | IS 6, 8 |
| 10. | $(\forall b)(\forall c)(\{b,c\} \in d \Rightarrow (b \in \cup d \wedge c \in \cup d))$ | UI 2 |
| 11. | $(\forall c)(\{\{e\},c\} \in d \Rightarrow (\{e\} \in \cup d \wedge c \in \cup d))$ | UI 10 |
| 12. | $\{\{e\},\{e,f\}\} \in d \Rightarrow (\{e\} \in \cup d \wedge \{e,f\} \in \cup d)$ | UI 11 |
| 13. | $\{e\} \in \cup d \wedge \{e,f\} \in \cup d$ | MP 9, 12 |
| 14. | $\{e,f\} \in \cup d$ | S 13 |
| 15. | $(\forall b)(\forall c)(\{b,c\} \in \cup d \Rightarrow (b \in \cup\cup d \wedge c \in \cup\cup d))$ | UI 2 |
| 16. | $(\forall c)(\{e,c\} \in \cup d \Rightarrow (e \in \cup\cup d \wedge c \in \cup\cup d))$ | UI 15 |
| 17. | $\{e,f\} \in \cup d \Rightarrow (e \in \cup\cup d \wedge f \in \cup\cup d)$ | UI 16 |
| 18. | $e \in \cup\cup d \wedge f \in \cup\cup d$ | MP 14, 17 |
| 19. | $(e,f) \in d \Rightarrow (e \in \cup\cup d \wedge f \in \cup\cup d)$ | CP 6-18 |
| 20. | $(\forall c)((e,c) \in d \Rightarrow (e \in \cup\cup d \wedge c \in \cup\cup d))$ | UG 5-19 |
| 21. | $(\forall b)(\forall c)((b,c) \in d \Rightarrow (b \in \cup\cup d \wedge c \in \cup\cup d))$ | UG 4-20 |
| 22. | $(\forall a)(\forall b)(\forall c)((b,c) \in a \Rightarrow (b \in \cup\cup a \wedge c \in \cup\cup a))$ | UG 3-21 |

□

**Theorem 2.100** $(\forall a)(\exists! b)(\forall c)(c \in b \Leftrightarrow (c \in \cup\cup a \wedge (\exists d)((c,d) \in a)))$

***Proof***

| | | |
|---|---|---|
| 1. | $(\forall a)(\forall e)(\exists b)(\forall c)(c \in b \Leftrightarrow (c \in a \wedge (\exists d)((c,d) \in e)))$ | AS 2.1 |
| 2. | $(\forall a)(\forall b)((\forall c)(c \in a \Leftrightarrow c \in b) \Rightarrow a = b)$ | Ax 2.1 |
| 3. | flag $f$ | FSUG |
| 4. | $(\forall e)(\exists b)(\forall c)(c \in b \Leftrightarrow (c \in \cup\cup f \wedge (\exists d)((c,d) \in e)))$ | UI 1 |
| 5. | $(\exists b)(\forall c)(c \in b \Leftrightarrow (c \in \cup\cup f \wedge (\exists d)((c,d) \in f)))$ | UI 4 |
| 6. | flag $g$ | FSUG |
| 7. | flag $h$ | FSUG |
| 8. | $(\forall c)(c \in g \Leftrightarrow (c \in \cup\cup f \wedge (\exists d)((c,d) \in f)))\wedge$ <br> $(\forall c)(c \in h \Leftrightarrow (c \in \cup\cup f \wedge (\exists d)((c,d) \in f)))$ | ACP |
| 9. | $(\forall c)(c \in g \Leftrightarrow (c \in \cup\cup f \wedge (\exists d)((c,d) \in f)))$ | S 8 |
| 10. | $(\forall c)(c \in h \Leftrightarrow (c \in \cup\cup f \wedge (\exists d)((c,d) \in f)))$ | S 8 |

| | | |
|---|---|---|
| 11. | flag $i$ | FSUG |
| 12. | $i \in g \Leftrightarrow (i \in \bigcup\bigcup f \wedge (\exists d)((i, d) \in f))$ | UI 9 |
| 13. | $i \in h \Leftrightarrow (i \in \bigcup\bigcup f \wedge (\exists d)((i, d) \in f))$ | UI 10 |
| 14. | $i \in g \Leftrightarrow i \in h$ | BT 12, 13 |
| 15. | $(\forall c)(c \in g \Leftrightarrow c \in h)$ | UG 11-14 |
| 16. | $(\forall b)((\forall c)(c \in g \Leftrightarrow c \in b) \Rightarrow g = b)$ | UI 2 |
| 17. | $(\forall c)(c \in g \Leftrightarrow c \in h) \Rightarrow g = h$ | UI 16 |
| 18. | $g = h$ | MP 15, 17 |
| 19. | $((\forall c)(c \in g \Leftrightarrow (c \in \bigcup\bigcup f \wedge (\exists d)((c, d) \in f)))\wedge$ $(\forall c)(c \in h \Leftrightarrow (c \in \bigcup\bigcup f \wedge (\exists d)((c, d) \in f)))) \Rightarrow g = h$ | CP 8-18 |
| 20. | $(\forall b)(((\forall c)(c \in g \Leftrightarrow (c \in \bigcup\bigcup f \wedge (\exists d)((c, d) \in f)))\wedge$ $(\forall c)(c \in b \Leftrightarrow (c \in \bigcup\bigcup f \wedge (\exists d)((c, d) \in f)))) \Rightarrow g = b)$ | UG 7-19 |
| 21. | $(\forall a)(\forall b)(((\forall c)(c \in a \Leftrightarrow (c \in \bigcup\bigcup f \wedge (\exists d)((c, d) \in f)))\wedge$ $(\forall c)(c \in b \Leftrightarrow (c \in \bigcup\bigcup f \wedge (\exists d)((c, d) \in f)))) \Rightarrow a = b)$ | UG 6-20 |
| 22. | $(\exists b)(\forall c)(c \in b \Leftrightarrow (c \in \bigcup\bigcup f \wedge (\exists d)((c, d) \in f)))\wedge$ $(\forall a)(\forall b)(((\forall c)(c \in a \Leftrightarrow (c \in \bigcup\bigcup f \wedge (\exists d)((c, d) \in f)))\wedge$ $(\forall c)(c \in b \Leftrightarrow (c \in \bigcup\bigcup f \wedge (\exists d)((c, d) \in f)))) \Rightarrow a = b)$ | C 5, 21 |
| 23. | $(\exists! b)(\forall c)(c \in b \Leftrightarrow (c \in \bigcup\bigcup f \wedge (\exists d)((c, d) \in f)))$ | UQ 22 |
| 24. | $(\forall a)(\exists! b)(\forall c)(c \in b \Leftrightarrow (c \in \bigcup\bigcup a \wedge (\exists d)((c, d) \in a)))$ | UG 3-23 |

□

**Axiom 2.23 (Definition of Predicate Which If Applied to the Argument $\alpha$ Gives the Atomic Formula "$\alpha$ is a function")**
$(\forall a)(a$ is a function $\Leftrightarrow$ $(a$ is a relation $\wedge (\forall b)(\forall c)(\forall d)(((b, c) \in a \wedge (b, d) \in a) \Rightarrow c = d)))$

**Theorem 2.101** $(\forall a)(a$ is a function $\Rightarrow (\exists! b)(\forall c)(c \in b \Leftrightarrow (c \in \bigcup\bigcup a \wedge (\exists d)((c, d) \in a))))$

***Proof***

| | | |
|---|---|---|
| 1. | $(\forall a)(\exists! b)(\forall c)(c \in b \Leftrightarrow (c \in \bigcup\bigcup a \wedge (\exists d)((c, d) \in a)))$ | T 2.100 |
| 2. | flag $e$ | FSUG |
| 3. | $(\exists! b)(\forall c)(c \in b \Leftrightarrow (c \in \bigcup\bigcup e \wedge (\exists d)((c, d) \in e)))$ | UI 1 |
| 4. | $\neg(e$ is a function$)\vee$ $(\exists! b)(\forall c)(c \in b \Leftrightarrow (c \in \bigcup\bigcup e \wedge (\exists d)((c, d) \in e)))$ | Ad 3 |
| 5. | $e$ is a function $\Rightarrow$ $(\exists! b)(\forall c)(c \in b \Leftrightarrow (c \in \bigcup\bigcup e \wedge (\exists d)((c, d) \in e)))$ | CE 4 |
| 6. | $(\forall a)(a$ is a function $\Rightarrow$ $(\exists! b)(\forall c)(c \in b \Leftrightarrow (c \in \bigcup\bigcup a \wedge (\exists d)((c, d) \in a))))$ | UG 2-5 |

□

**Axiom 2.24 (Definitions of Operations, from Theorem 2.101, Which If Applied to the Argument $\alpha$ Give the Terms "D($\alpha$)" and "domain of $\alpha$")**
$(\forall a)(a \text{ is a function} \Rightarrow (\forall b)(b \in \mathrm{D}(a) \Leftrightarrow (b \in \cup\cup a \wedge (\exists c)((b, c) \in a))))$
$(\forall a)(a \text{ is a function} \Rightarrow (\forall b)(b \in \text{domain of } a \Leftrightarrow (b \in \cup\cup a \wedge (\exists c)((b, c) \in a))))$

**Theorem 2.102** $(\forall a)(\exists! b)(\forall c)(c \in b \Leftrightarrow (c \in \cup\cup a \wedge (\exists d)((d, c) \in a)))$

***Proof***

| | | |
|---|---|---|
| 1. | $(\forall a)(\forall e)(\exists b)(\forall c)(c \in b \Leftrightarrow (c \in a \wedge (\exists d)((d, c) \in e)))$ | AS 2.1 |
| 2. | $(\forall a)(\forall b)((\forall c)(c \in a \Leftrightarrow c \in b) \Rightarrow a = b)$ | Ax 2.1 |
| 3. | flag $f$ | FSUG |
| 4. | $(\forall e)(\exists b)(\forall c)(c \in b \Leftrightarrow (c \in \cup\cup f \wedge (\exists d)((d, c) \in e)))$ | UI 1 |
| 5. | $(\exists b)(\forall c)(c \in b \Leftrightarrow (c \in \cup\cup f \wedge (\exists d)((d, c) \in f)))$ | UI 4 |
| 6. | flag $g$ | FSUG |
| 7. | flag $h$ | FSUG |
| 8. | $(\forall c)(c \in g \Leftrightarrow (c \in \cup\cup f \wedge (\exists d)((d, c) \in f)))\wedge$ $(\forall c)(c \in h \Leftrightarrow (c \in \cup\cup f \wedge (\exists d)((d, c) \in f)))$ | ACP |
| 9. | $(\forall c)(c \in g \Leftrightarrow (c \in \cup\cup f \wedge (\exists d)((d, c) \in f)))$ | S 8 |
| 10. | $(\forall c)(c \in h \Leftrightarrow (c \in \cup\cup f \wedge (\exists d)((d, c) \in f)))$ | S 8 |
| 11. | flag $i$ | FSUG |
| 12. | $i \in g \Leftrightarrow (i \in \cup\cup f \wedge (\exists d)((d, i) \in f))$ | UI 9 |
| 13. | $i \in h \Leftrightarrow (i \in \cup\cup f \wedge (\exists d)((d, i) \in f))$ | UI 10 |
| 14. | $i \in g \Leftrightarrow i \in h$ | BT 12, 13 |
| 15. | $(\forall c)(c \in g \Leftrightarrow c \in h)$ | UG 11-14 |
| 16. | $(\forall b)((\forall c)(c \in g \Leftrightarrow c \in b) \Rightarrow g = b)$ | UI 2 |
| 17. | $(\forall c)(c \in g \Leftrightarrow c \in h) \Rightarrow g = h$ | UI 16 |
| 18. | $g = h$ | MP 15, 17 |
| 19. | $((\forall c)(c \in g \Leftrightarrow (c \in \cup\cup f \wedge (\exists d)((d, c) \in f)))\wedge$ $(\forall c)(c \in h \Leftrightarrow (c \in \cup\cup f \wedge (\exists d)((d, c) \in f)))) \Rightarrow g = h$ | CP 8-18 |
| 20. | $(\forall b)(((\forall c)(c \in g \Leftrightarrow (c \in \cup\cup f \wedge (\exists d)((d, c) \in f)))\wedge$ $(\forall c)(c \in b \Leftrightarrow (c \in \cup\cup f \wedge (\exists d)((d, c) \in f)))) \Rightarrow g = b)$ | UG 7-19 |
| 21. | $(\forall a)(\forall b)(((\forall c)(c \in a \Leftrightarrow (c \in \cup\cup f \wedge (\exists d)((d, c) \in f)))\wedge$ $(\forall c)(c \in b \Leftrightarrow (c \in \cup\cup f \wedge (\exists d)((d, c) \in f)))) \Rightarrow a = b)$ | UG 6-20 |
| 22. | $(\exists b)(\forall c)(c \in b \Leftrightarrow (c \in \cup\cup f \wedge (\exists d)((d, c) \in f)))\wedge$ $(\forall a)(\forall b)(((\forall c)(c \in a \Leftrightarrow (c \in \cup\cup f \wedge (\exists d)((d, c) \in f)))\wedge$ $(\forall c)(c \in b \Leftrightarrow (c \in \cup\cup f \wedge (\exists d)((d, c) \in f)))) \Rightarrow a = b)$ | C 5, 21 |
| 23. | $(\exists! b)(\forall c)(c \in b \Leftrightarrow (c \in \cup\cup f \wedge (\exists d)((d, c) \in f)))$ | UQ 22 |
| 24. | $(\forall a)(\exists! b)(\forall c)(c \in b \Leftrightarrow (c \in \cup\cup a \wedge (\exists d)((d, c) \in a)))$ | UG 3-23 |

□

**Theorem 2.103** $(\forall a)(a \text{ is a function} \Rightarrow (\exists! b)(\forall c)(c \in b \Leftrightarrow (c \in \cup\cup a \wedge (\exists d)((d, c) \in a))))$

*Proof*

| | | |
|---|---|---|
| 1. | $(\forall a)(\exists! b)(\forall c)(c \in b \Leftrightarrow (c \in \cup\cup a \wedge (\exists d)((d, c) \in a)))$ | T 2.102 |
| 2. | flag $e$ | FSUG |
| 3. | $(\exists! b)(\forall c)(c \in b \Leftrightarrow (c \in \cup\cup e \wedge (\exists d)((d, c) \in e)))$ | UI 1 |
| 4. | $\neg(e$ is a function$)\vee$ $(\exists! b)(\forall c)(c \in b \Leftrightarrow (c \in \cup\cup e \wedge (\exists d)((d, c) \in e)))$ | Ad 3 |
| 5. | $e$ is a function $\Rightarrow$ $(\exists! b)(\forall c)(c \in b \Leftrightarrow (c \in \cup\cup e \wedge (\exists d)((d, c) \in e)))$ | CE 4 |
| 6. | $(\forall a)(a$ is a function $\Rightarrow$ $(\exists! b)(\forall c)(c \in b \Leftrightarrow (c \in \cup\cup a \wedge (\exists d)((d, c) \in a))))$ | UG 2-5 |

□

**Axiom 2.25 (Definitions of Operations, from Theorem 2.103, Which If Applied to the Argument $\alpha$ Give the Terms "R($\alpha$)" and "range of $\alpha$")**
$(\forall a)(a$ is a function $\Rightarrow (\forall b)(b \in \mathrm{R}(a) \Leftrightarrow (b \in \cup\cup a \wedge (\exists c)((c, b) \in a))))$
$(\forall a)(a$ is a function $\Rightarrow (\forall b)(b \in$ range of $a \Leftrightarrow (b \in \cup\cup a \wedge (\exists c)((c, b) \in a))))$

**Theorem 2.104** $(\forall a)(\forall b)(\forall c)((a$ is a function $\wedge (b, c) \in a) \Rightarrow b \in \mathrm{D}(a))$

*Proof*

| | | |
|---|---|---|
| 1. | $(\forall a)(\forall b)(\forall c)((b, c) \in a \Rightarrow (b \in \cup\cup a \wedge c \in \cup\cup a))$ | T 2.99 |
| 2. | $(\forall a)(a$ is a function $\Rightarrow$ $(\forall b)(b \in \mathrm{D}(a) \Leftrightarrow (b \in \cup\cup a \wedge (\exists c)((b, c) \in a))))$ | Ax 2.24 |
| 3. | flag $d$ | FSUG |
| 4. | flag $e$ | FSUG |
| 5. | flag $f$ | FSUG |
| 6. | $d$ is a function $\wedge (e, f) \in d$ | ACP |
| 7. | $(e, f) \in d$ | S 6 |
| 8. | $(\forall b)(\forall c)((b, c) \in d \Rightarrow (b \in \cup\cup d \wedge c \in \cup\cup d))$ | UI 1 |
| 9. | $(\forall c)((e, c) \in d \Rightarrow (e \in \cup\cup d \wedge c \in \cup\cup d))$ | UI 8 |
| 10. | $(e, f) \in d \Rightarrow (e \in \cup\cup d \wedge f \in \cup\cup d)$ | UI 9 |
| 11. | $e \in \cup\cup d \wedge f \in \cup\cup d$ | MP 7, 10 |
| 12. | $e \in \cup\cup d$ | S 11 |
| 13. | $(\exists c)((e, c) \in d)$ | EG 7 |
| 14. | $e \in \cup\cup d \wedge (\exists c)((e, c) \in d)$ | C 12, 13 |
| 15. | $d$ is a function | S 6 |
| 16. | $d$ is a function $\Rightarrow$ $(\forall b)(b \in \mathrm{D}(d) \Leftrightarrow (b \in \cup\cup d \wedge (\exists c)((b, c) \in d)))$ | UI 2 |
| 17. | $(\forall b)(b \in \mathrm{D}(d) \Leftrightarrow (b \in \cup\cup d \wedge (\exists c)((b, c) \in d)))$ | MP 15, 16 |

| | | |
|---|---|---|
| 18. | $e \in \mathrm{D}(d) \Leftrightarrow (e \in \cup\cup d \wedge (\exists c)((e, c) \in d))$ | UI 17 |
| 19. | $e \in \mathrm{D}(d)$ | BEl 14, 18 |
| 20. | $(d \text{ is a function} \wedge (e, f) \in d) \Rightarrow e \in \mathrm{D}(d)$ | CP 6-19 |
| 21. | $(\forall c)((d \text{ is a function} \wedge (e, c) \in d) \Rightarrow e \in \mathrm{D}(d))$ | UG 5-20 |
| 22. | $(\forall b)(\forall c)((d \text{ is a function} \wedge (b, c) \in d) \Rightarrow b \in \mathrm{D}(d))$ | UG 4-21 |
| 23. | $(\forall a)(\forall b)(\forall c)((a \text{ is a function} \wedge (b, c) \in a) \Rightarrow b \in \mathrm{D}(a))$ | UG 3-22 |

□

**Theorem 2.105** $(\forall a)(\forall b)(\forall c)((a \text{ is a function} \wedge (b, c) \in a) \Rightarrow c \in \mathrm{R}(a))$

***Proof***

| | | |
|---|---|---|
| 1. | $(\forall a)(\forall b)(\forall c)((b, c) \in a \Rightarrow (b \in \cup\cup a \wedge c \in \cup\cup a))$ | T 2.99 |
| 2. | $(\forall a)(a \text{ is a function} \Rightarrow$ <br> $(\forall b)(b \in \mathrm{R}(a) \Leftrightarrow (b \in \cup\cup a \wedge (\exists c)((c, b) \in a))))$ | Ax 2.25 |
| 3. | flag $d$ | FSUG |
| 4. | flag $e$ | FSUG |
| 5. | flag $f$ | FSUG |
| 6. | $d \text{ is a function} \wedge (e, f) \in d$ | ACP |
| 7. | $(e, f) \in d$ | S 6 |
| 8. | $(\forall b)(\forall c)((b, c) \in d \Rightarrow (b \in \cup\cup d \wedge c \in \cup\cup d))$ | UI 1 |
| 9. | $(\forall c)((e, c) \in d \Rightarrow (e \in \cup\cup d \wedge c \in \cup\cup d))$ | UI 8 |
| 10. | $(e, f) \in d \Rightarrow (e \in \cup\cup d \wedge f \in \cup\cup d)$ | UI 9 |
| 11. | $e \in \cup\cup d \wedge f \in \cup\cup d$ | MP 7, 10 |
| 12. | $f \in \cup\cup d$ | S 11 |
| 13. | $(\exists c)((c, f) \in d)$ | EG 7 |
| 14. | $f \in \cup\cup d \wedge (\exists c)((c, f) \in d)$ | C 12, 13 |
| 15. | $d \text{ is a function} \Rightarrow$ <br> $(\forall b)(b \in \mathrm{R}(d) \Leftrightarrow (b \in \cup\cup d \wedge (\exists c)((c, b) \in d)))$ | UI 2 |
| 16. | $d$ is a function | S 6 |
| 17. | $(\forall b)(b \in \mathrm{R}(d) \Leftrightarrow (b \in \cup\cup d \wedge (\exists c)((c, b) \in d)))$ | MP 15, 16 |
| 18. | $f \in \mathrm{R}(d) \Leftrightarrow (f \in \cup\cup d \wedge (\exists c)((c, f) \in d))$ | UI 17 |
| 19. | $f \in \mathrm{R}(d)$ | BEl 14, 18 |
| 20. | $(d \text{ is a function} \wedge (e, f) \in d) \Rightarrow f \in \mathrm{R}(d)$ | CP 6-19 |
| 21. | $(\forall c)((d \text{ is a function} \wedge (e, c) \in d) \Rightarrow c \in \mathrm{R}(d))$ | UG 5-20 |
| 22. | $(\forall b)(\forall c)((d \text{ is a function} \wedge (b, c) \in d) \Rightarrow c \in \mathrm{R}(d))$ | UG 4-21 |
| 23. | $(\forall a)(\forall b)(\forall c)((a \text{ is a function} \wedge (b, c) \in a) \Rightarrow c \in \mathrm{R}(a))$ | UG 3-22 |

□

**Theorem 2.106** $(\forall a)(\forall b)(\forall c)((a \text{ is a function} \wedge a \subseteq b \times c) \Rightarrow \mathrm{D}(a) \subseteq b)$

***Proof***

| | | |
|---|---|---|
| 1. | $(\forall a)(a \text{ is a function} \Rightarrow (\forall b)(b \in \mathrm{D}(a) \Leftrightarrow (b \in \cup\cup a \wedge (\exists c)((b, c) \in a))))$ | Ax 2.24 |
| 2. | $(\forall a)(\forall b)(a \subseteq b \Leftrightarrow (\forall c)(c \in a \Rightarrow c \in b))$ | Ax 2.9 |
| 3. | $(\forall a)(\forall b)(\forall c)(\forall d)((c \in a \wedge d \in b) \Leftrightarrow (c, d) \in a \times b)$ | T 2.92 |
| 4. | flag $e$ | FSUG |
| 5. | flag $f$ | FSUG |
| 6. | flag $g$ | FSUG |
| 7. | $e$ is a function $\wedge\, e \subseteq f \times g$ | ACP |
| 8. | $e$ is a function | S 7 |
| 9. | $e \subseteq f \times g$ | S 7 |
| 10. | flag $h$ | FSUG |
| 11. | $h \in \mathrm{D}(e)$ | ACP |
| 12. | $e \text{ is a function} \Rightarrow (\forall b)(b \in \mathrm{D}(e) \Leftrightarrow (b \in \cup\cup e \wedge (\exists c)((b, c) \in e)))$ | UI 1 |
| 13. | $(\forall b)(b \in \mathrm{D}(e) \Leftrightarrow (b \in \cup\cup e \wedge (\exists c)((b, c) \in e)))$ | MP 8, 12 |
| 14. | $h \in \mathrm{D}(e) \Leftrightarrow (h \in \cup\cup e \wedge (\exists c)((h, c) \in e))$ | UI 13 |
| 15. | $h \in \cup\cup e \wedge (\exists c)((h, c) \in e)$ | BEl 11, 14 |
| 16. | $(\exists c)((h, c) \in e)$ | S 15 |
| 17. | $(h, i) \in e$ | EI 16, flag $i$ |
| 18. | $(\forall b)(e \subseteq b \Leftrightarrow (\forall c)(c \in e \Rightarrow c \in b))$ | UI 2 |
| 19. | $e \subseteq f \times g \Leftrightarrow (\forall c)(c \in e \Rightarrow c \in f \times g)$ | UI 18 |
| 20. | $(\forall c)(c \in e \Rightarrow c \in f \times g)$ | BEl 9, 19 |
| 21. | $(h, i) \in e \Rightarrow (h, i) \in f \times g$ | UI 20 |
| 22. | $(h, i) \in f \times g$ | MP 17, 21 |
| 23. | $(\forall b)(\forall c)(\forall d)((c \in f \wedge d \in b) \Leftrightarrow (c, d) \in f \times b)$ | UI 3 |
| 24. | $(\forall c)(\forall d)((c \in f \wedge d \in g) \Leftrightarrow (c, d) \in f \times g)$ | UI 23 |
| 25. | $(\forall d)((h \in f \wedge d \in g) \Leftrightarrow (h, d) \in f \times g)$ | UI 24 |
| 26. | $(h \in f \wedge i \in g) \Leftrightarrow (h, i) \in f \times g$ | UI 25 |
| 27. | $h \in f \wedge i \in g$ | BEl 22, 26 |
| 28. | $h \in f$ | S 27 |
| 29. | $h \in \mathrm{D}(e) \Rightarrow h \in f$ | CP 11-28 |
| 30. | $(\forall c)(c \in \mathrm{D}(e) \Rightarrow c \in f)$ | UG 10-29 |
| 31. | $(\forall b)(\mathrm{D}(e) \subseteq b \Leftrightarrow (\forall c)(c \in \mathrm{D}(e) \Rightarrow c \in b))$ | UI 2 |
| 32. | $\mathrm{D}(e) \subseteq f \Leftrightarrow (\forall c)(c \in \mathrm{D}(e) \Rightarrow c \in f)$ | UI 31 |
| 33. | $\mathrm{D}(e) \subseteq f$ | BEl 30, 32 |
| 34. | $(e \text{ is a function} \wedge e \subseteq f \times g) \Rightarrow \mathrm{D}(e) \subseteq f$ | CP 7-33 |
| 35. | $(\forall c)((e \text{ is a function} \wedge e \subseteq f \times c) \Rightarrow \mathrm{D}(e) \subseteq f)$ | UG 6-34 |
| 36. | $(\forall b)(\forall c)((e \text{ is a function} \wedge e \subseteq b \times c) \Rightarrow \mathrm{D}(e) \subseteq b)$ | UG 5-35 |
| 37. | $(\forall a)(\forall b)(\forall c)((a \text{ is a function} \wedge a \subseteq b \times c) \Rightarrow \mathrm{D}(a) \subseteq b)$ | UG 4-36 |

□

**Theorem 2.107** $(\forall a)(\forall b)(\forall c)((a \text{ is a function} \wedge a \subseteq b \times c) \Rightarrow \mathrm{R}(a) \subseteq c)$

***Proof***

| | | |
|---|---|---|
| 1. | $(\forall a)(a \text{ is a function} \Rightarrow$ $(\forall b)(b \in \mathrm{R}(a) \Leftrightarrow (b \in \cup\cup a \wedge (\exists c)((c, b) \in a))))$ | Ax 2.25 |
| 2. | $(\forall a)(\forall b)(a \subseteq b \Leftrightarrow (\forall c)(c \in a \Rightarrow c \in b))$ | Ax 2.9 |
| 3. | $(\forall a)(\forall b)(\forall c)(\forall d)((c \in a \wedge d \in b) \Leftrightarrow (c, d) \in a \times b)$ | T 2.92 |
| 4. | flag $e$ | FSUG |
| 5. | flag $f$ | FSUG |
| 6. | flag $g$ | FSUG |
| 7. | $e$ is a function $\wedge\, e \subseteq f \times g$ | ACP |
| 8. | $e$ is a function | S 7 |
| 9. | $e \subseteq f \times g$ | S 7 |
| 10. | flag $h$ | FSUG |
| 11. | $h \in \mathrm{R}(e)$ | ACP |
| 12. | $e \text{ is a function} \Rightarrow (\forall b)(b \in \mathrm{R}(e) \Leftrightarrow$ $(b \in \cup\cup e \wedge (\exists c)((c, b) \in e)))$ | UI 1 |
| 13. | $(\forall b)(b \in \mathrm{R}(e) \Leftrightarrow$ $(b \in \cup\cup e \wedge (\exists c)((c, b) \in e)))$ | MP 8, 12 |
| 14. | $h \in \mathrm{R}(e) \Leftrightarrow (h \in \cup\cup e \wedge (\exists c)((c, h) \in e))$ | UI 13 |
| 15. | $h \in \cup\cup e \wedge (\exists c)((c, h) \in e)$ | BEl 11, 14 |
| 16. | $(\exists c)((c, h) \in e)$ | S 15 |
| 17. | $(i, h) \in e$ | EI 16, flag $i$ |
| 18. | $(\forall b)(e \subseteq b \Leftrightarrow (\forall c)(c \in e \Rightarrow c \in b))$ | UI 2 |
| 19. | $e \subseteq f \times g \Leftrightarrow (\forall c)(c \in e \Rightarrow c \in f \times g)$ | UI 18 |
| 20. | $(\forall c)(c \in e \Rightarrow c \in f \times g)$ | BEl 9, 19 |
| 21. | $(i, h) \in e \Rightarrow (i, h) \in f \times g$ | UI 20 |
| 22. | $(i, h) \in f \times g$ | MP 17, 21 |
| 23. | $(\forall b)(\forall c)(\forall d)((c \in f \wedge d \in b) \Leftrightarrow$ $(c, d) \in f \times b)$ | UI 3 |
| 24. | $(\forall c)(\forall d)((c \in f \wedge d \in g) \Leftrightarrow (c, d) \in f \times g)$ | UI 23 |
| 25. | $(\forall d)((i \in f \wedge d \in g) \Leftrightarrow (i, d) \in f \times g)$ | UI 24 |
| 26. | $(i \in f \wedge h \in g) \Leftrightarrow (i, h) \in f \times g$ | UI 25 |
| 27. | $i \in f \wedge h \in g$ | BEl 22, 26 |
| 28. | $h \in g$ | S 27 |
| 29. | $h \in \mathrm{R}(e) \Rightarrow h \in g$ | CP 11-28 |
| 30. | $(\forall c)(c \in \mathrm{R}(e) \Rightarrow c \in g)$ | UG 10-29 |
| 31. | $(\forall b)(\mathrm{R}(e) \subseteq b \Leftrightarrow (\forall c)(c \in \mathrm{R}(e) \Rightarrow c \in b))$ | UI 2 |
| 32. | $\mathrm{R}(e) \subseteq g \Leftrightarrow (\forall c)(c \in \mathrm{R}(e) \Rightarrow c \in g)$ | UI 31 |
| 33. | $\mathrm{R}(e) \subseteq g$ | BEl 30, 32 |
| 34. | $(e \text{ is a function} \wedge e \subseteq f \times g) \Rightarrow \mathrm{R}(e) \subseteq g$ | CP 7-33 |
| 35. | $(\forall c)((e \text{ is a function} \wedge e \subseteq f \times c) \Rightarrow \mathrm{R}(e) \subseteq c)$ | UG 6-34 |
| 36. | $(\forall b)(\forall c)((e \text{ is a function} \wedge e \subseteq b \times c) \Rightarrow \mathrm{R}(e) \subseteq c)$ | UG 5-35 |
| 37. | $(\forall a)(\forall b)(\forall c)((a \text{ is a function} \wedge a \subseteq b \times c) \Rightarrow \mathrm{R}(a) \subseteq c)$ | UG 4-36 |

□

**Theorem 2.108** $(\forall a)(\forall b)((a \text{ is a function} \wedge b \in \mathrm{D}(a)) \Rightarrow (\exists! c)((b, c) \in a))$

***Proof***

| | | |
|---|---|---|
| 1. | $(\forall a)(a \text{ is a function} \Rightarrow$ $(\forall b)(b \in \cup\cup a \Leftrightarrow (b \in \cup\cup a \wedge (\exists c)((b, c) \in a))))$ | Ax 2.24 |
| 2. | $(\forall a)(a \text{ is a function} \Leftrightarrow (a \text{ is a relation} \wedge$ $(\forall b)(\forall c)(\forall d)(((b, c) \in a \wedge (b, d) \in a) \Rightarrow c = d)))$ | Ax 2.23 |
| 3. | flag $e$ | FSUG |
| 4. | flag $f$ | FSUG |
| 5. | $e \text{ is a function} \wedge f \in \mathrm{D}(e)$ | ACP |
| 6. | $e$ is a function | S 5 |
| 7. | $e \text{ is a function} \Rightarrow$ $(\forall b)(b \in \mathrm{D}(e) \Leftrightarrow (b \in \cup\cup e \wedge (\exists c)((b, c) \in e)))$ | UI 1 |
| 8. | $(\forall b)(b \in \mathrm{D}(e) \Leftrightarrow (b \in \cup\cup e \wedge (\exists c)((b, c) \in e)))$ | MP 6, 7 |
| 9. | $f \in \mathrm{D}(e)$ | S 5 |
| 10. | $f \in \mathrm{D}(e) \Leftrightarrow (f \in \cup\cup e \wedge (\exists c)((f, c) \in e))$ | UI 8 |
| 11. | $f \in \cup\cup e \wedge (\exists c)((f, c) \in e)$ | BEl 9, 10 |
| 12. | $(\exists c)((f, c) \in e)$ | S 11 |
| 13. | $(f, g) \in e$ | EI 12, flag $g$ |
| 14. | flag $h$ | FSUG |
| 15. | $(f, h) \in e$ | ACP |
| 16. | $(f, h) \in e \wedge (f, g) \in e$ | C 13, 15 |
| 17. | $e \text{ is a function} \Leftrightarrow (e \text{ is a relation} \wedge$ $(\forall b)(\forall c)(\forall d)(((b, c) \in e \wedge (b, d) \in e)$ $\Rightarrow c = d))$ | UI 2 |
| 18. | $e \text{ is a relation} \wedge$ $(\forall b)(\forall c)(\forall d)(((b, c) \in e \wedge (b, d) \in e) \Rightarrow c = d)$ | BEl 6, 17 |
| 19. | $(\forall b)(\forall c)(\forall d)(((b, c) \in e \wedge (b, d) \in e) \Rightarrow c = d)$ | S 18 |
| 20. | $(\forall c)(\forall d)(((f, c) \in e \wedge (f, d) \in e) \Rightarrow c = d)$ | UI 19 |
| 21. | $(\forall d)(((f, h) \in e \wedge (f, d) \in e) \Rightarrow h = d)$ | UI 20 |
| 22. | $((f, h) \in e \wedge (f, g) \in e) \Rightarrow h = g$ | UI 21 |
| 23. | $h = g$ | MP 16, 22 |
| 24. | $(f, h) \in e \Rightarrow h = g$ | CP 15-23 |
| 25. | $(\forall d)((f, d) \in e \Rightarrow d = g)$ | UG 14-24 |
| 26. | $(f, g) \in e \wedge (\forall d)((f, d) \in e \Rightarrow d = g)$ | C 13, 25 |
| 27. | $(\exists c)((f, c) \in e \wedge (\forall d)((f, d) \in e \Rightarrow d = c))$ | EG 26 |
| 28. | $(\exists! c)((f, c) \in e)$ | UQ 27 |
| 29. | $(e \text{ is a function} \wedge f \in \mathrm{D}(e)) \Rightarrow (\exists! c)((f, c) \in e)$ | CP 5-28 |
| 30. | $(\forall b)((e \text{ is a function} \wedge b \in \mathrm{D}(e)) \Rightarrow (\exists! c)((b, c) \in e))$ | UG 4-29 |
| 31. | $(\forall a)(\forall b)((a \text{ is a function} \wedge b \in \mathrm{D}(a)) \Rightarrow (\exists! c)((b, c) \in a))$ | UG 3-30 |

□

**Axiom 2.26 (Definition of Operation, from Theorem 2.108, Which If Applied to the Arguments $\alpha$ and $\beta$ Gives the Term "$\alpha(\beta)$")**
$(\forall a)(\forall b)((a \text{ is a function} \wedge b \in \mathrm{D}(a)) \Rightarrow (b, a(b)) \in a)$

**Axiom 2.27 (Definitions of Predicates Which If Applied to the Arguments $\alpha$, $\beta$ and $\gamma$ Give the Atomic Formulas "$\alpha : \beta \to \gamma$" and "$\alpha$ is a function from $\beta$ into $\gamma$")**

$(\forall a)(\forall b)(\forall c)(a : b \to c \Leftrightarrow ((a \text{ is a function} \wedge \mathrm{D}(a) = b) \wedge \mathrm{R}(a) \subseteq c))$
$(\forall a)(\forall b)(\forall c)(a \text{ is a function from } b \text{ into } c \Leftrightarrow ((a \text{ is a function} \wedge \mathrm{D}(a) = b) \wedge \mathrm{R}(a) \subseteq c))$

## 2.12 Inductive Set

**Axiom 2.28 (Zermelo Axiom: Axiom of Infinity)** $(\exists a)(\phi \in a \wedge (\forall b)(b \in a \Rightarrow b \cup \{b\} \in a))$

**Axiom 2.29 (Definitions of Predicates Which If Applied to the Argument $\alpha$ Give the Atomic Formulas "$\alpha$ is an inductive set" and "$\alpha$ is inductive")**

$(\forall a)(a \text{ is an inductive set} \Leftrightarrow (\phi \in a \wedge (\forall b)(b \in a \Rightarrow b \cup \{b\} \in a)))$
$(\forall a)(a \text{ is inductive} \Leftrightarrow (\phi \in a \wedge (\forall b)(b \in a \Rightarrow b \cup \{b\} \in a)))$

**Theorem 2.109** $(\exists a)(a \text{ is inductive})$

***Proof***

| | | |
|---|---|---|
| 1. | $(\exists a)(\phi \in a \wedge (\forall b)(b \in a \Rightarrow b \cup \{b\} \in a))$ | Ax 2.28 |
| 2. | $(\forall a)(a \text{ is inductive} \Leftrightarrow (\phi \in a \wedge (\forall b)(b \in a \Rightarrow b \cup \{b\} \in a)))$ | Ax 2.29 |
| 3. | $\phi \in c \wedge (\forall b)(b \in c \Rightarrow b \cup \{b\} \in c)$ | EI 1, flag $c$ |
| 4. | $c \text{ is inductive} \Leftrightarrow (\phi \in c \wedge (\forall b)(b \in c \Rightarrow b \cup \{b\} \in c))$ | UI 2 |
| 5. | $c$ is inductive | BEI 3, 4 |
| 6. | $(\exists a)(a \text{ is inductive})$ | EG 5 |

□

**Theorem 2.110** $(\forall a)(\forall b)((a \text{ is inductive} \wedge b \text{ is inductive}) \Rightarrow a \cap b \text{ is inductive})$

***Proof***

| | | |
|---|---|---|
| 1. | $(\forall a)(a \text{ is inductive} \Leftrightarrow (\phi \in a \wedge (\forall b)(b \in a \Rightarrow b \cup \{b\} \in a)))$ | Ax 2.29 |
| 2. | $(\forall a)(\forall b)(\forall c)(c \in a \cap b \Leftrightarrow (c \in a \wedge c \in b))$ | Ax 2.15 |
| 3. | flag $d$ | FSUG |
| 4. | flag $e$ | FSUG |
| 5. | $d$ is inductive $\wedge$ $e$ is inductive | ACP |
| 6. | $d$ is inductive | S 5 |
| 7. | $d$ is inductive $\Leftrightarrow$ $(\phi \in d \wedge (\forall b)(b \in d \Rightarrow b \cup \{b\} \in d))$ | UI 1 |
| 8. | $\phi \in d \wedge (\forall b)(b \in d \Rightarrow b \cup \{b\} \in d)$ | BEl 6, 7 |
| 9. | $\phi \in d$ | S 8 |
| 10. | $(\forall b)(b \in d \Rightarrow b \cup \{b\} \in d)$ | S 8 |
| 11. | $e$ is inductive | S 5 |
| 12. | $e$ is inductive $\Leftrightarrow$ $(\phi \in e \wedge (\forall b)(b \in e \Rightarrow b \cup \{b\} \in e))$ | UI 1 |
| 13. | $\phi \in e \wedge (\forall b)(b \in e \Rightarrow b \cup \{b\} \in e)$ | BEl 11, 12 |
| 14. | $\phi \in e$ | S 13 |
| 15. | $(\forall b)(b \in e \Rightarrow b \cup \{b\} \in e)$ | S 13 |
| 16. | $\phi \in d \wedge \phi \in e$ | C 9, 14 |
| 17. | $(\forall b)(\forall c)(c \in d \cap b \Leftrightarrow (c \in d \wedge c \in b))$ | UI 2 |
| 18. | $(\forall c)(c \in d \cap e \Leftrightarrow (c \in d \wedge c \in e))$ | UI 17 |
| 19. | $\phi \in d \cap e \Leftrightarrow (\phi \in d \wedge \phi \in e)$ | UI 18 |
| 20. | $\phi \in d \cap e$ | BEl 16, 19 |
| 21. | flag $f$ | FSUG |
| 22. | $f \in d \cap e$ | ACP |
| 23. | $f \in d \cap e \Leftrightarrow (f \in d \wedge f \in e)$ | UI 18 |
| 24. | $f \in d \wedge f \in e$ | BEl 22, 23 |
| 25. | $f \in d$ | S 24 |
| 26. | $f \in d \Rightarrow f \cup \{f\} \in d$ | UI 10 |
| 27. | $f \cup \{f\} \in d$ | MP 25, 26 |
| 28. | $f \in e$ | S 24 |
| 29. | $f \in e \Rightarrow f \cup \{f\} \in e$ | UI 15 |
| 30. | $f \cup \{f\} \in e$ | MP 28, 29 |
| 31. | $f \cup \{f\} \in d \wedge f \cup \{f\} \in e$ | C 27, 30 |
| 32. | $f \cup \{f\} \in d \cap e \Leftrightarrow (f \cup \{f\} \in d \wedge f \cup \{f\} \in e)$ | UI 18 |
| 33. | $f \cup \{f\} \in d \cap e$ | BEl 31, 32 |
| 34. | $f \in d \cap e \Rightarrow f \cup \{f\} \in d \cap e$ | CP 22-33 |
| 35. | $(\forall b)(b \in d \cap e \Rightarrow b \cup \{b\} \in d \cap e)$ | UG 21-34 |
| 36. | $\phi \in d \cap e \wedge (\forall b)(b \in d \cap e \Rightarrow b \cup \{b\} \in d \cap e)$ | C 20, 35 |
| 37. | $d \cap e$ is inductive $\Leftrightarrow$ $(\phi \in d \cap e \wedge (\forall b)(b \in d \cap e \Rightarrow b \cup \{b\} \in d \cap e))$ | UI 1 |
| 38. | $d \cap e$ is inductive | BEl 36, 37 |

| | | |
|---|---|---|
| 39. | ($d$ is inductive $\wedge$ $e$ is inductive) $\Rightarrow$ $d \cap e$ is inductive | CP 5-38 |
| 40. | $(\forall b)((d$ is inductive $\wedge$ $b$ is inductive$) \Rightarrow$ $d \cap b$ is inductive$)$ | UG 4-39 |
| 41. | $(\forall a)(\forall b)((a$ is inductive $\wedge$ $b$ is inductive$) \Rightarrow$ $a \cap b$ is inductive$)$ | UG 3-40 |

□

**Theorem 2.111** $(\exists a)(\forall b)(b \in a \Leftrightarrow (\forall c)(c$ is inductive $\Rightarrow b \in c))$

***Proof***

| | | |
|---|---|---|
| 1. | $(\exists a)(a$ is inductive$)$ | T 2.109 |
| 2. | $(\forall a)(a \in \mathrm{P}(a))$ | T 2.80 |
| 3. | $(\forall a)(\exists b)(\forall c)(c \in b \Leftrightarrow (c \in a \wedge c$ is inductive$))$ | AS 2.1 |
| 4. | $(\forall a)((\exists b)(b \in a) \Rightarrow a \neq \phi)$ | T 2.48 |
| 5. | $(\forall a)(a \neq \phi \Rightarrow (\forall b)(b \in \cap a \Leftrightarrow (\forall c)(c \in a \Rightarrow b \in c)))$ | Ax 2.16 |
| 6. | $(\forall a)(\forall b)((a$ is inductive $\wedge$ $b$ is inductive$) \Rightarrow$ $a \cap b$ is inductive$)$ | T 2.110 |
| 7. | $(\forall a)(\forall b)(a \cap b \subseteq a)$ | T 2.55 |
| 8. | $(\forall a)(\forall b)(b \in \mathrm{P}(a) \Leftrightarrow b \subseteq a)$ | Ax 2.19 |
| 9. | $(\forall a)(\forall b)(\forall c)(c \in a \cap b \Leftrightarrow (c \in a \wedge c \in b))$ | Ax 2.15 |
| 10. | $d$ is inductive | EI 1, flag $d$ |
| 11. | $d \in \mathrm{P}(d)$ | UI 2 |
| 12. | $d \in \mathrm{P}(d) \wedge d$ is inductive | C 10, 11 |
| 13. | $(\exists b)(\forall c)(c \in b \Leftrightarrow (c \in \mathrm{P}(d) \wedge c$ is inductive$))$ | UI 3 |
| 14. | $(\forall c)(c \in e \Leftrightarrow (c \in \mathrm{P}(d) \wedge c$ is inductive$))$ | EI 13, flag $e$ |
| 15. | $d \in e \Leftrightarrow (d \in \mathrm{P}(d) \wedge d$ is inductive$)$ | UI 14 |
| 16. | $d \in e$ | BEI 12, 15 |
| 17. | $(\exists b)(b \in e)$ | EG 16 |
| 18. | $(\exists b)(b \in e) \Rightarrow e \neq \phi$ | UI 4 |
| 19. | $e \neq \phi$ | MP 17, 18 |
| 20. | $e \neq \phi \Rightarrow (\forall b)(b \in \cap e \Leftrightarrow (\forall c)(c \in e \Rightarrow b \in c))$ | UI 5 |
| 21. | $(\forall b)(b \in \cap e \Leftrightarrow (\forall c)(c \in e \Rightarrow b \in c))$ | MP 19, 20 |
| 22. | flag $f$ | FSUG |
| 23. | $(\forall c)(c$ is inductive $\Rightarrow f \in c)$ | ACP |
| 24. | flag $g$ | FSUG |
| 25. | $g$ is inductive $\Rightarrow f \in g$ | UI 23 |
| 26. | $g \in \mathrm{P}(d) \wedge g$ is inductive | ACP |
| 27. | $g$ is inductive | S 26 |
| 28. | $f \in g$ | MP 25, 27 |
| 29. | $(g \in \mathrm{P}(d) \wedge g$ is inductive$) \Rightarrow f \in g$ | CP 26-28 |

| | | |
|---|---|---|
| 30. | $g \in e$ | ACP |
| 31. | $g \in e \Leftrightarrow (g \in \mathrm{P}(d) \wedge g$ is inductive) | UI 14 |
| 32. | $g \in \mathrm{P}(d) \wedge g$ is inductive | BEl 30, 31 |
| 33. | $f \in g$ | MP 29, 32 |
| 34. | $g \in e \Rightarrow f \in g$ | CP 30-33 |
| 35. | $(\forall c)(c \in e \Rightarrow f \in c)$ | UG 24-34 |
| 36. | $f \in \cap e \Leftrightarrow (\forall c)(c \in e \Rightarrow f \in c)$ | UI 21 |
| 37. | $f \in \cap e$ | BEl 35, 36 |
| 38. | $(\forall c)(c$ is inductive $\Rightarrow f \in c) \Rightarrow f \in \cap e$ | CP 23-37 |
| 39. | $f \in \cap e$ | ACP |
| 40. | flag $h$ | FSUG |
| 41. | $h$ is inductive | ACP |
| 42. | $d$ is inductive $\wedge$ $h$ is inductive | C 10, 41 |
| 43. | $(\forall b)((d$ is inductive $\wedge$ $b$ is inductive) $\Rightarrow$ $d \cap b$ is inductive) | UI 6 |
| 44. | ($d$ is inductive $\wedge$ $h$ is inductive) $\Rightarrow$ $d \cap h$ is inductive | UI 43 |
| 45. | $d \cap h$ is inductive | MP 42, 44 |
| 46. | $(\forall b)(d \cap b \subseteq d)$ | UI 7 |
| 47. | $d \cap h \subseteq d$ | UI 46 |
| 48. | $(\forall b)(b \in \mathrm{P}(d) \Leftrightarrow b \subseteq d)$ | UI 8 |
| 49. | $d \cap h \in \mathrm{P}(d) \Leftrightarrow d \cap h \subseteq d$ | UI 48 |
| 50. | $d \cap h \in \mathrm{P}(d)$ | BEl 47, 49 |
| 51. | $d \cap h \in \mathrm{P}(d) \wedge d \cap h$ is inductive | C 45, 50 |
| 52. | $d \cap h \in e \Leftrightarrow (d \cap h \in \mathrm{P}(d) \wedge d \cap h$ is inductive) | UI 14 |
| 53. | $d \cap h \in e$ | BEl 51, 52 |
| 54. | $f \in \cap e \Leftrightarrow (\forall c)(c \in e \Rightarrow f \in c)$ | UI 21 |
| 55. | $(\forall c)(c \in e \Rightarrow f \in c)$ | BEl 39, 54 |
| 56. | $d \cap h \in e \Rightarrow f \in d \cap h$ | UI 55 |
| 57. | $f \in d \cap h$ | MP 53, 56 |
| 58. | $(\forall b)(\forall c)(c \in d \cap b \Leftrightarrow (c \in d \wedge c \in b))$ | UI 9 |
| 59. | $(\forall c)(c \in d \cap h \Leftrightarrow (c \in d \wedge c \in h))$ | UI 58 |
| 60. | $f \in d \cap h \Leftrightarrow (f \in d \wedge f \in h)$ | UI 59 |
| 61. | $f \in d \wedge f \in h$ | BEl 57, 60 |
| 62. | $f \in h$ | S 61 |
| 63. | $h$ is inductive $\Rightarrow f \in h$ | CP 41-62 |
| 64. | $(\forall c)(c$ is inductive $\Rightarrow f \in c)$ | UG 40-63 |
| 65. | $f \in \cap e \Rightarrow (\forall c)(c$ is inductive $\Rightarrow f \in c)$ | CP 39-64 |
| 66. | $((\forall c)(c$ is inductive $\Rightarrow f \in c) \Rightarrow f \in \cap e) \wedge$ $(f \in \cap e \Rightarrow (\forall c)(c$ is inductive $\Rightarrow f \in c))$ | C 38, 65 |
| 67. | $f \in \cap e \Leftrightarrow (\forall c)(c$ is inductive $\Rightarrow f \in c)$ | BE 66 |
| 68. | $(\forall b)(b \in \cap e \Leftrightarrow (\forall c)(c$ is inductive $\Rightarrow b \in c))$ | UG 22-67 |
| 69. | $(\exists a)(\forall b)(b \in a \Leftrightarrow (\forall c)(c$ is inductive $\Rightarrow b \in c))$ | EG 68 |

□

**Theorem 2.112** $(\forall a)(\forall b)(((\forall c)(c \in a \Leftrightarrow (\forall d)(d \text{ is inductive} \Rightarrow c \in d)) \wedge (\forall c)(c \in b \Leftrightarrow (\forall d)(d \text{ is inductive} \Rightarrow c \in d))) \Rightarrow a = b)$

***Proof***

| | | |
|---|---|---|
| 1. | $(\forall a)(\forall b)((\forall c)(c \in a \Leftrightarrow c \in b) \Rightarrow a = b)$ | Ax 2.1 |
| 2. | flag $e$ | FSUG |
| 3. | flag $f$ | FSUG |
| 4. | $(\forall c)(c \in e \Leftrightarrow (\forall d)(d \text{ is inductive} \Rightarrow c \in d)) \wedge$ $(\forall c)(c \in f \Leftrightarrow (\forall d)(d \text{ is inductive} \Rightarrow c \in d))$ | ACP |
| 5. | $(\forall c)(c \in e \Leftrightarrow (\forall d)(d \text{ is inductive} \Rightarrow c \in d))$ | S 4 |
| 6. | $(\forall c)(c \in f \Leftrightarrow (\forall d)(d \text{ is inductive} \Rightarrow c \in d))$ | S 4 |
| 7. | flag $g$ | FSUG |
| 8. | $g \in e \Leftrightarrow (\forall d)(d \text{ is inductive} \Rightarrow g \in d)$ | UI 5 |
| 9. | $g \in f \Leftrightarrow (\forall d)(d \text{ is inductive} \Rightarrow g \in d)$ | UI 6 |
| 10. | $g \in e \Leftrightarrow g \in f$ | BT 8, 9 |
| 11. | $(\forall c)(c \in e \Leftrightarrow c \in f)$ | UG 7-10 |
| 12. | $(\forall b)((\forall c)(c \in e \Leftrightarrow c \in b) \Rightarrow e = b)$ | UI 1 |
| 13. | $(\forall c)(c \in e \Leftrightarrow c \in f) \Rightarrow e = f$ | UI 12 |
| 14. | $e = f$ | MP 11, 13 |
| 15. | $((\forall c)(c \in e \Leftrightarrow (\forall d)(d \text{ is inductive} \Rightarrow c \in d)) \wedge$ $(\forall c)(c \in f \Leftrightarrow (\forall d)(d \text{ is inductive} \Rightarrow c \in d))) \Rightarrow e = f$ | CP 4-14 |
| 16. | $(\forall b)(((\forall c)(c \in e \Leftrightarrow (\forall d)(d \text{ is inductive} \Rightarrow c \in d)) \wedge$ $(\forall c)(c \in b \Leftrightarrow (\forall d)(d \text{ is inductive} \Rightarrow c \in d))) \Rightarrow e = b)$ | UG 3-15 |
| 17. | $(\forall a)(\forall b)(((\forall c)(c \in a \Leftrightarrow (\forall d)(d \text{ is inductive} \Rightarrow c \in d)) \wedge$ $(\forall c)(c \in b \Leftrightarrow (\forall d)(d \text{ is inductive} \Rightarrow c \in d))) \Rightarrow a = b)$ | UG 2-16 |

□

**Theorem 2.113** $(\exists! a)(\forall b)(b \in a \Leftrightarrow (\forall c)(c \text{ is inductive} \Rightarrow b \in c))$

***Proof***

| | | |
|---|---|---|
| 1. | $(\exists a)(\forall b)(b \in a \Leftrightarrow (\forall c)(c \text{ is inductive} \Rightarrow b \in c))$ | T 2.111 |
| 2. | $(\forall d)(\forall e)(((\forall b)(b \in d \Leftrightarrow (\forall c)(c \text{ is inductive} \Rightarrow b \in c)) \wedge$ $(\forall b)(b \in e \Leftrightarrow (\forall c)(c \text{ is inductive} \Rightarrow b \in c))) \Rightarrow d = e)$ | T 2.112 |
| 3. | $(\exists a)(\forall b)(b \in a \Leftrightarrow (\forall c)(c \text{ is inductive} \Rightarrow b \in c)) \wedge$ $(\forall d)(\forall e)(((\forall b)(b \in d \Leftrightarrow (\forall c)(c \text{ is inductive} \Rightarrow b \in c)) \wedge$ $(\forall b)(b \in e \Leftrightarrow (\forall c)(c \text{ is inductive} \Rightarrow b \in c))) \Rightarrow d = e)$ | C 1, 2 |
| 4. | $(\exists! a)(\forall b)(b \in a \Leftrightarrow (\forall c)(c \text{ is inductive} \Rightarrow b \in c))$ | UQ 3 |

□

**Axiom 2.30 (Definitions of Constants "set of non-negative integers" and "N" from Theorem 2.113)**

$(\forall a)(a \in \text{set of non-negative integers} \Leftrightarrow (\forall b)(b \text{ is inductive} \Rightarrow a \in b))$
$(\forall a)(a \in \text{N} \Leftrightarrow (\forall b)(b \text{ is inductive} \Rightarrow a \in b))$

**Theorem 2.114** N is inductive

***Proof***

| | | |
|---|---|---|
| 1. | $(\forall a)(a \text{ is inductive} \Leftrightarrow (\phi \in a \wedge (\forall b)(b \in a \Rightarrow b \cup \{b\} \in a)))$ | Ax 2.29 |
| 2. | $(\forall a)(a \in \text{N} \Leftrightarrow (\forall b)(b \text{ is inductive} \Rightarrow a \in b))$ | Ax 2.30 |
| 3. | flag $c$ | FSUG |
| 4. | $c$ is inductive | ACP |
| 5. | $c \text{ is inductive} \Leftrightarrow (\phi \in c \wedge (\forall b)(b \in c \Rightarrow b \cup \{b\} \in c))$ | UI 1 |
| 6. | $\phi \in c \wedge (\forall b)(b \in c \Rightarrow b \cup \{b\} \in c)$ | BEl 4, 5 |
| 7. | $\phi \in c$ | S 6 |
| 8. | $c \text{ is inductive} \Rightarrow \phi \in c$ | CP 4-7 |
| 9. | $(\forall b)(b \text{ is inductive} \Rightarrow \phi \in b)$ | UG 3-8 |
| 10. | $\phi \in \text{N} \Leftrightarrow (\forall b)(b \text{ is inductive} \Rightarrow \phi \in b)$ | UI 2 |
| 11. | $\phi \in \text{N}$ | BEl 9, 10 |
| 12. | flag $d$ | FSUG |
| 13. | $d \in \text{N}$ | ACP |
| 14. | $d \in \text{N} \Leftrightarrow (\forall b)(b \text{ is inductive} \Rightarrow d \in b)$ | UI 2 |
| 15. | $(\forall b)(b \text{ is inductive} \Rightarrow d \in b)$ | BEl 13, 14 |
| 16. | flag $e$ | FSUG |
| 17. | $e \text{ is inductive} \Rightarrow d \in e$ | UI 15 |
| 18. | $e$ is inductive | ACP |
| 19. | $d \in e$ | MP 17, 18 |
| 20. | $e \text{ is inductive} \Leftrightarrow (\phi \in e \wedge (\forall b)(b \in e \Rightarrow b \cup \{b\} \in e))$ | UI 1 |
| 21. | $\phi \in e \wedge (\forall b)(b \in e \Rightarrow b \cup \{b\} \in e)$ | BEl 18, 20 |
| 22. | $(\forall b)(b \in e \Rightarrow b \cup \{b\} \in e)$ | S 21 |
| 23. | $d \in e \Rightarrow d \cup \{d\} \in e$ | UI 22 |
| 24. | $d \cup \{d\} \in e$ | MP 19, 23 |
| 25. | $e \text{ is inductive} \Rightarrow d \cup \{d\} \in e$ | CP 18-24 |
| 26. | $(\forall b)(b \text{ is inductive} \Rightarrow d \cup \{d\} \in b)$ | UG 16-25 |
| 27. | $d \cup \{d\} \in \text{N} \Leftrightarrow (\forall b)(b \text{ is inductive} \Rightarrow d \cup \{d\} \in b)$ | UI 2 |
| 28. | $d \cup \{d\} \in \text{N}$ | BEl 26, 27 |
| 29. | $d \in \text{N} \Rightarrow d \cup \{d\} \in \text{N}$ | CP 13-28 |
| 30. | $(\forall b)(b \in \text{N} \Rightarrow b \cup \{b\} \in \text{N})$ | UG 12-29 |
| 31. | $\phi \in \text{N} \wedge (\forall b)(b \in \text{N} \Rightarrow b \cup \{b\} \in \text{N})$ | C 11, 30 |
| 32. | $\text{N is inductive} \Leftrightarrow (\phi \in \text{N} \wedge (\forall b)(b \in \text{N} \Rightarrow b \cup \{b\} \in \text{N}))$ | UI 1 |
| 33. | N is inductive | BEl 31, 32 |

□

**Theorem 2.115** $\phi \in \mathrm{N}$

***Proof***

| | | |
|---|---|---|
| 1. | $(\forall a)(a \text{ is inductive} \Leftrightarrow (\phi \in a \wedge (\forall b)(b \in a \Rightarrow b \cup \{b\} \in a)))$ | Ax 2.29 |
| 2. | N is inductive | T 2.114 |
| 3. | $\mathrm{N} \text{ is inductive} \Leftrightarrow (\phi \in \mathrm{N} \wedge (\forall b)(b \in \mathrm{N} \Rightarrow b \cup \{b\} \in \mathrm{N}))$ | UI 1 |
| 4. | $\phi \in \mathrm{N} \wedge (\forall b)(b \in \mathrm{N} \Rightarrow b \cup \{b\} \in \mathrm{N})$ | BEl 2, 3 |
| 5. | $\phi \in \mathrm{N}$ | S 4 |

□

**Theorem 2.116** $(\forall a)(a \in \mathrm{N} \Rightarrow a \cup \{a\} \in \mathrm{N})$

***Proof***

| | | |
|---|---|---|
| 1. | $(\forall b)(b \text{ is inductive} \Leftrightarrow (\phi \in b \wedge (\forall a)(a \in b \Rightarrow a \cup \{a\} \in b)))$ | Ax 2.29 |
| 2. | N is inductive | T 2.114 |
| 3. | $\mathrm{N} \text{ is inductive} \Leftrightarrow (\phi \in \mathrm{N} \wedge (\forall a)(a \in \mathrm{N} \Rightarrow a \cup \{a\} \in \mathrm{N}))$ | UI 1 |
| 4. | $\phi \in \mathrm{N} \wedge (\forall a)(a \in \mathrm{N} \Rightarrow a \cup \{a\} \in \mathrm{N})$ | BEl 2, 3 |
| 5. | $(\forall a)(a \in \mathrm{N} \Rightarrow a \cup \{a\} \in \mathrm{N})$ | S 4 |

□

**Theorem 2.117** $(\forall a)(a \text{ is inductive} \Rightarrow \mathrm{N} \subseteq a)$

***Proof***

| | | |
|---|---|---|
| 1. | $(\forall a)(a \in \mathrm{N} \Leftrightarrow (\forall b)(b \text{ is inductive} \Rightarrow a \in b))$ | Ax 2.30 |
| 2. | $(\forall a)(\forall b)(a \subseteq b \Leftrightarrow (\forall c)(c \in a \Rightarrow c \in b))$ | Ax 2.9 |
| 3. | flag $d$ | FSUG |
| 4. | $d$ is inductive | ACP |
| 5. | flag $e$ | FSUG |
| 6. | $e \in \mathrm{N}$ | ACP |
| 7. | $e \in \mathrm{N} \Leftrightarrow (\forall b)(b \text{ is inductive} \Rightarrow e \in b)$ | UI 1 |
| 8. | $(\forall b)(b \text{ is inductive} \Rightarrow e \in b)$ | BEl 6, 7 |
| 9. | $d \text{ is inductive} \Rightarrow e \in d$ | UI 8 |
| 10. | $e \in d$ | MP 4, 9 |
| 11. | $e \in \mathrm{N} \Rightarrow e \in d$ | CP 6-10 |
| 12. | $(\forall c)(c \in \mathrm{N} \Rightarrow c \in d)$ | UG 5-11 |
| 13. | $(\forall b)(\mathrm{N} \subseteq b \Leftrightarrow (\forall c)(c \in \mathrm{N} \Rightarrow c \in b))$ | UI 2 |
| 14. | $\mathrm{N} \subseteq d \Leftrightarrow (\forall c)(c \in \mathrm{N} \Rightarrow c \in d)$ | UI 13 |
| 15. | $\mathrm{N} \subseteq d$ | BEl 12, 14 |
| 16. | $d \text{ is inductive} \Rightarrow \mathrm{N} \subseteq d$ | CP 4-15 |
| 17. | $(\forall a)(a \text{ is inductive} \Rightarrow \mathrm{N} \subseteq a)$ | UG 3-16 |

□

**Theorem 2.118** $(\forall a)(((a \subseteq \mathrm{N} \wedge \phi \in a) \wedge (\forall b)(b \in a \Rightarrow b \cup \{b\} \in a)) \Rightarrow a = \mathrm{N})$

***Proof***

| | | |
|---|---|---|
| 1. | $(\forall a)(a \text{ is inductive} \Leftrightarrow (\phi \in a \wedge (\forall b)(b \in a \Rightarrow b \cup \{b\} \in a)))$ | Ax 2.29 |
| 2. | $(\forall a)(a \text{ is inductive} \Rightarrow \mathrm{N} \subseteq a)$ | T 2.117 |
| 3. | $(\forall a)(\forall b)(a = b \Leftrightarrow (a \subseteq b \wedge b \subseteq a))$ | T 2.24 |
| 4. | flag $c$ | FSUG |
| 5. | $(c \subseteq \mathrm{N} \wedge \phi \in c) \wedge (\forall b)(b \in c \Rightarrow b \cup \{b\} \in c)$ | ACP |
| 6. | $c \subseteq \mathrm{N} \wedge \phi \in c$ | S 5 |
| 7. | $(\forall b)(b \in c \Rightarrow b \cup \{b\} \in c)$ | S 5 |
| 8. | $c \subseteq \mathrm{N}$ | S 6 |
| 9. | $\phi \in c$ | S 6 |
| 10. | $\phi \in c \wedge (\forall b)(b \in c \Rightarrow b \cup \{b\} \in c)$ | C 7, 9 |
| 11. | $c \text{ is inductive} \Leftrightarrow (\phi \in c \wedge (\forall b)(b \in c \Rightarrow b \cup \{b\} \in c))$ | UI 1 |
| 12. | $c$ is inductive | BEl 10, 11 |
| 13. | $c \text{ is inductive} \Rightarrow \mathrm{N} \subseteq c$ | UI 2 |
| 14. | $\mathrm{N} \subseteq c$ | MP 12, 13 |
| 15. | $c \subseteq \mathrm{N} \wedge \mathrm{N} \subseteq c$ | C 8, 14 |
| 16. | $(\forall b)(c = b \Leftrightarrow (c \subseteq b \wedge b \subseteq c))$ | UI 3 |
| 17. | $c = \mathrm{N} \Leftrightarrow (c \subseteq \mathrm{N} \wedge \mathrm{N} \subseteq c)$ | UI 16 |
| 18. | $c = \mathrm{N}$ | BEl 15, 17 |
| 19. | $((c \subseteq \mathrm{N} \wedge \phi \in c) \wedge (\forall b)(b \in c \Rightarrow b \cup \{b\} \in c)) \Rightarrow c = \mathrm{N}$ | CP 5-18 |
| 20. | $(\forall a)(((a \subseteq \mathrm{N} \wedge \phi \in a) \wedge (\forall b)(b \in a \Rightarrow b \cup \{b\} \in a)) \Rightarrow a = \mathrm{N})$ | UG 4-19 |

□

## 2.13 Transitive Set

**Axiom 2.31 (Definitions of Predicates Which If Applied to the Argument $\alpha$ Give the Atomic Formulas "$\alpha$ is a transitive set" and "$\alpha$ is transitive")**

$(\forall a)(a \text{ is a transitive set} \Leftrightarrow (\forall b)(\forall c)((b \in c \wedge c \in a) \Rightarrow b \in a))$
$(\forall a)(a \text{ is transitive} \Leftrightarrow (\forall b)(\forall c)((b \in c \wedge c \in a) \Rightarrow b \in a))$

**Theorem 2.119** $(\forall a)(a \text{ is transitive} \Leftrightarrow \cup a \subseteq a)$

***Proof***

| | | |
|---|---|---|
| 1. | $(\forall a)(a \text{ is transitive} \Leftrightarrow (\forall b)(\forall c)((b \in c \wedge c \in a) \Rightarrow b \in a))$ | Ax 2.31 |
| 2. | $(\forall a)(\forall b)(b \in \cup a \Leftrightarrow (\exists c)(b \in c \wedge c \in a))$ | Ax 2.7 |
| 3. | $(\forall a)(\forall b)(a \subseteq b \Leftrightarrow (\forall c)(c \in a \Rightarrow c \in b))$ | Ax 2.9 |
| 4. | flag $d$ | FSUG |
| 5. | $d$ is transitive | ACP |
| 6. | $d \text{ is transitive} \Leftrightarrow (\forall b)(\forall c)((b \in c \wedge c \in d) \Rightarrow b \in d)$ | UI 1 |
| 7. | $(\forall b)(\forall c)((b \in c \wedge c \in d) \Rightarrow b \in d)$ | BEl 5, 6 |
| 8. | flag $e$ | FSUG |
| 9. | $e \in \cup d$ | ACP |
| 10. | $(\forall b)(b \in \cup d \Leftrightarrow (\exists c)(b \in c \wedge c \in d))$ | UI 2 |
| 11. | $e \in \cup d \Leftrightarrow (\exists c)(e \in c \wedge c \in d)$ | UI 10 |
| 12. | $(\exists c)(e \in c \wedge c \in d)$ | BEl 9, 11 |
| 13. | $e \in f \wedge f \in d$ | EI 12, flag $f$ |
| 14. | $(\forall c)((e \in c \wedge c \in d) \Rightarrow e \in d)$ | UI 7 |
| 15. | $(e \in f \wedge f \in d) \Rightarrow e \in d$ | UI 14 |
| 16. | $e \in d$ | MP 13, 15 |
| 17. | $e \in \cup d \Rightarrow e \in d$ | CP 9-16 |
| 18. | $(\forall c)(c \in \cup d \Rightarrow c \in d)$ | UG 8-17 |
| 19. | $(\forall b)(\cup d \subseteq b \Leftrightarrow (\forall c)(c \in \cup d \Rightarrow c \in b))$ | UI 3 |
| 20. | $\cup d \subseteq d \Leftrightarrow (\forall c)(c \in \cup d \Rightarrow c \in d)$ | UI 19 |
| 21. | $\cup d \subseteq d$ | BEl 18, 20 |
| 22. | $d \text{ is transitive} \Rightarrow \cup d \subseteq d$ | CP 5-21 |
| 23. | $\cup d \subseteq d$ | ACP |
| 24. | $(\forall b)(\cup d \subseteq b \Leftrightarrow (\forall c)(c \in \cup d \Rightarrow c \in b))$ | UI 3 |
| 25. | $\cup d \subseteq d \Leftrightarrow (\forall c)(c \in \cup d \Rightarrow c \in d)$ | UI 24 |
| 26. | $(\forall c)(c \in \cup d \Rightarrow c \in d)$ | BEl 23, 25 |
| 27. | flag $g$ | FSUG |
| 28. | $g \in \cup d \Rightarrow g \in d$ | UI 26 |
| 29. | flag $h$ | FSUG |
| 30. | $g \in h \wedge h \in d$ | ACP |
| 31. | $(\exists c)(g \in c \wedge c \in d)$ | EG 30 |
| 32. | $(\forall b)(b \in \cup d \Leftrightarrow (\exists c)(b \in c \wedge c \in d))$ | UI 2 |
| 33. | $g \in \cup d \Leftrightarrow (\exists c)(g \in c \wedge c \in d)$ | UI 32 |
| 34. | $g \in \cup d$ | BEl 31, 33 |
| 35. | $g \in d$ | MP 28, 34 |
| 36. | $(g \in h \wedge h \in d) \Rightarrow g \in d$ | CP 30-35 |
| 37. | $(\forall c)((g \in c \wedge c \in d) \Rightarrow g \in d)$ | UG 29-36 |
| 38. | $(\forall b)(\forall c)((b \in c \wedge c \in d) \Rightarrow b \in d)$ | UG 27-37 |
| 39. | $d \text{ is transitive} \Leftrightarrow (\forall b)(\forall c)((b \in c \wedge c \in d) \Rightarrow b \in d)$ | UI 1 |
| 40. | $d$ is transitive | BEl 38, 39 |

| | | |
|---|---|---|
| 41. | $\cup d \subseteq d \Rightarrow d$ is transitive | CP 23-40 |
| 42. | ($d$ is transitive $\Rightarrow \cup d \subseteq d) \wedge (\cup d \subseteq d \Rightarrow d$ is transitive) | C 22, 41 |
| 43. | $d$ is transitive $\Leftrightarrow \cup d \subseteq d$ | BE 42 |
| 44. | $(\forall a)(a$ is transitive $\Leftrightarrow \cup a \subseteq a)$ | UG 4-43 |

□

**Theorem 2.120** $\phi$ is transitive

***Proof***

| | | |
|---|---|---|
| 1. | $(\forall a)(a \notin \phi)$ | Ax 2.14 |
| 2. | $(\forall a)(\forall b)(a \notin b \Leftrightarrow \neg(a \in b))$ | Ax 2.12 |
| 3. | $(\forall a)(a$ is transitive $\Leftrightarrow (\forall b)(\forall c)((b \in c \wedge c \in a) \Rightarrow b \in a))$ | Ax 2.31 |
| 4. | flag $d$ | FSUG |
| 5. | flag $e$ | FSUG |
| 6. | $d \in e \wedge e \in \phi$ | AIP |
| 7. | $e \in \phi$ | S 6 |
| 8. | $e \notin \phi$ | UI 1 |
| 9. | $(\forall b)(e \notin b \Leftrightarrow \neg(e \in b))$ | UI 2 |
| 10. | $e \notin \phi \Leftrightarrow \neg(e \in \phi)$ | UI 9 |
| 11. | $\neg(e \in \phi)$ | BEl 8, 10 |
| 12. | $e \in \phi \wedge \neg(e \in \phi)$ | C 7, 11 |
| 13. | $\neg(d \in e \wedge e \in \phi)$ | IP 6-12 |
| 14. | $(\neg(d \in e \wedge e \in \phi)) \vee d \in \phi$ | Ad 13 |
| 15. | $(d \in e \wedge e \in \phi) \Rightarrow d \in \phi$ | CE 14 |
| 16. | $(\forall c)((d \in c \wedge c \in \phi) \Rightarrow d \in \phi)$ | UG 5-15 |
| 17. | $(\forall b)(\forall c)((b \in c \wedge c \in \phi) \Rightarrow b \in \phi)$ | UG 4-16 |
| 18. | $\phi$ is transitive $\Leftrightarrow (\forall b)(\forall c)((b \in c \wedge c \in \phi) \Rightarrow b \in \phi)$ | UI 3 |
| 19. | $\phi$ is transitive | BEl 17, 18 |

□

**Theorem 2.121** $(\forall a)(a \in \mathrm{N} \Rightarrow a$ is transitive)

***Proof***

| | | |
|---|---|---|
| 1. | $\phi \in \mathrm{N}$ | T 2.115 |
| 2. | $\phi$ is transitive | T 2.120 |
| 3. | $(\forall a)(\exists b)(\forall c)(c \in b \Leftrightarrow (c \in a \wedge c$ is transitive$))$ | AS 2.1 |
| 4. | $(\forall a)(a \in \mathrm{N} \Rightarrow a \cup \{a\} \in \mathrm{N})$ | T 2.116 |
| 5. | $(\forall a)(\forall b)(\forall c)(c \in a \cup b \Leftrightarrow (c \in a \vee c \in b))$ | T 2.18 |
| 6. | $(\forall a)(\forall b)(b \in \{a\} \Leftrightarrow b = a)$ | T 2.8 |
| 7. | $(\forall a)(a$ is transitive $\Leftrightarrow (\forall b)(\forall c)((b \in c \wedge c \in a) \Rightarrow b \in a))$ | Ax 2.31 |
| 8. | $(\forall a)(\forall b)(\forall c)(c \in a \Rightarrow c \in a \cup b)$ | T 2.28 |
| 9. | $(\forall a)(\forall b)(a \subseteq b \Leftrightarrow (\forall c)(c \in a \Rightarrow c \in b))$ | Ax 2.9 |
| 10. | $(\forall a)(((a \subseteq \mathrm{N} \wedge \phi \in a) \wedge (\forall b)(b \in a \Rightarrow b \cup \{b\} \in a)) \Rightarrow a = \mathrm{N})$ | T 2.118 |
| 11. | $\phi \in \mathrm{N} \wedge \phi$ is transitive | C 1, 2 |
| 12. | $(\exists b)(\forall c)(c \in b \Leftrightarrow (c \in \mathrm{N} \wedge c$ is transitive$))$ | UI 3 |
| 13. | $(\forall c)(c \in d \Leftrightarrow (c \in \mathrm{N} \wedge c$ is transitive$))$ | EI 12, flag $d$ |
| 14. | $\phi \in d \Leftrightarrow (\phi \in \mathrm{N} \wedge \phi$ is transitive$)$ | UI 13 |
| 15. | $\phi \in d$ | BEl 11, 14 |
| 16. | flag $e$ | FSUG |
| 17. | $e \in d$ | ACP |
| 18. | $e \in d \Leftrightarrow (e \in \mathrm{N} \wedge e$ is transitive$)$ | UI 13 |
| 19. | $e \in \mathrm{N} \wedge e$ is transitive | BEl 17, 18 |
| 20. | $e \in \mathrm{N}$ | S 19 |
| 21. | $e \in \mathrm{N} \Rightarrow e \cup \{e\} \in \mathrm{N}$ | UI 4 |
| 22. | $e \cup \{e\} \in \mathrm{N}$ | MP 20, 21 |
| 23. | flag $f$ | FSUG |
| 24. | flag $g$ | FSUG |
| 25. | $f \in g \wedge g \in e \cup \{e\}$ | ACP |
| 26. | $g \in e \cup \{e\}$ | S 25 |
| 27. | $(\forall b)(\forall c)(c \in e \cup b \Leftrightarrow (c \in e \vee c \in b))$ | UI 5 |
| 28. | $(\forall c)(c \in e \cup \{e\} \Leftrightarrow (c \in e \vee c \in \{e\}))$ | UI 27 |
| 29. | $g \in e \cup \{e\} \Leftrightarrow (g \in e \vee g \in \{e\})$ | UI 28 |
| 30. | $g \in e \vee g \in \{e\}$ | BEl 26, 29 |
| 31. | $(\forall b)(b \in \{e\} \Leftrightarrow b = e)$ | UI 6 |
| 32. | $g \in \{e\} \Leftrightarrow g = e$ | UI 31 |
| 33. | $g \in e \vee g = e$ | D 30, 32 |
| 34. | $f \in g$ | S 25 |
| 35. | $e$ is transitive | S 19 |
| 36. | $g \in e$ | ACP |
| 37. | $f \in g \wedge g \in e$ | C 34, 36 |
| 38. | $e$ is transitive $\Leftrightarrow (\forall b)(\forall c)((b \in c \wedge c \in e) \Rightarrow b \in e)$ | UI 7 |
| 39. | $(\forall b)(\forall c)((b \in c \wedge c \in e) \Rightarrow b \in e)$ | BEl 35, 38 |
| 40. | $(\forall c)((f \in c \wedge c \in e) \Rightarrow f \in e)$ | UI 39 |

| | | |
|---|---|---|
| 41. | $(f \in g \land g \in e) \Rightarrow f \in e$ | UI 40 |
| 42. | $f \in e$ | MP 37, 41 |
| 43. | $g \in e \Rightarrow f \in e$ | CP 36-42 |
| 44. | $g = e$ | ACP |
| 45. | $f \in e$ | IS 34, 44 |
| 46. | $g = e \Rightarrow f \in e$ | CP 44-45 |
| 47. | $f \in e \lor f \in e$ | D 33, 43, 46 |
| 48. | $f \in e$ | Du 47 |
| 49. | $(\forall b)(\forall c)(c \in e \Rightarrow c \in e \cup b)$ | UI 8 |
| 50. | $(\forall c)(c \in e \Rightarrow c \in e \cup \{e\})$ | UI 49 |
| 51. | $f \in e \Rightarrow f \in e \cup \{e\}$ | UI 50 |
| 52. | $f \in e \cup \{e\}$ | MP 48, 51 |
| 53. | $(f \in g \land g \in e \cup \{e\}) \Rightarrow f \in e \cup \{e\}$ | CP 25-52 |
| 54. | $(\forall c)((f \in c \land c \in e \cup \{e\}) \Rightarrow f \in e \cup \{e\})$ | UG 24-53 |
| 55. | $(\forall b)(\forall c)((b \in c \land c \in e \cup \{e\}) \Rightarrow b \in e \cup \{e\})$ | UG 23-54 |
| 56. | $e \cup \{e\}$ is transitive $\Leftrightarrow$ | |
| | $(\forall b)(\forall c)((b \in c \land c \in e \cup \{e\}) \Rightarrow b \in e \cup \{e\})$ | UI 7 |
| 57. | $e \cup \{e\}$ is transitive | BEl 55, 56 |
| 58. | $e \cup \{e\} \in \mathrm{N} \land e \cup \{e\}$ is transitive | C 22, 57 |
| 59. | $e \cup \{e\} \in d \Leftrightarrow (e \cup \{e\} \in \mathrm{N} \land e \cup \{e\}$ is transitive) | UI 13 |
| 60. | $e \cup \{e\} \in d$ | BEl 58, 59 |
| 61. | $e \in d \Rightarrow e \cup \{e\} \in d$ | CP 17- 60 |
| 62. | $(\forall b)(b \in d \Rightarrow b \cup \{b\} \in d)$ | UG 16-61 |
| 63. | flag $h$ | FSUG |
| 64. | $h \in d$ | ACP |
| 65. | $h \in d \Leftrightarrow (h \in \mathrm{N} \land h$ is transitive) | UI 13 |
| 66. | $h \in \mathrm{N} \land h$ is transitive | BEl 64, 65 |
| 67. | $h \in \mathrm{N}$ | S 66 |
| 68. | $h \in d \Rightarrow h \in \mathrm{N}$ | CP 64-67 |
| 69. | $(\forall c)(c \in d \Rightarrow c \in \mathrm{N})$ | UG 63-68 |
| 70. | $(\forall b)(d \subseteq b \Leftrightarrow (\forall c)(c \in d \Rightarrow c \in b))$ | UI 9 |
| 71. | $d \subseteq \mathrm{N} \Leftrightarrow (\forall c)(c \in d \Rightarrow c \in \mathrm{N})$ | UI 70 |
| 72. | $d \subseteq \mathrm{N}$ | BEl 69, 71 |
| 73. | $d \subseteq \mathrm{N} \land \phi \in d$ | C 15, 72 |
| 74. | $(d \subseteq \mathrm{N} \land \phi \in d) \land (\forall b)(b \in d \Rightarrow b \cup \{b\} \in d)$ | C 62, 73 |
| 75. | $((d \subseteq \mathrm{N} \land \phi \in d) \land (\forall b)(b \in d \Rightarrow b \cup \{b\} \in d)) \Rightarrow d = \mathrm{N}$ | UI 10 |
| 76. | $d = \mathrm{N}$ | MP 74, 75 |
| 77. | flag $i$ | FSUG |
| 78. | $i \in d$ | ACP |
| 79. | $i \in d \Leftrightarrow (i \in \mathrm{N} \land i$ is transitive) | UI 13 |
| 80. | $i \in \mathrm{N} \land i$ is transitive | BEl 78, 79 |

| | | |
|---|---|---|
| 81. | $i$ is transitive | S 80 |
| 82. | $i \in d \Rightarrow i$ is transitive | CP 78-81 |
| 83. | $i \in \mathrm{N} \Rightarrow i$ is transitive | IS 76, 82 |
| 84. | $(\forall a)(a \in \mathrm{N} \Rightarrow a$ is transitive) | UG 77-83 |

□

**Theorem 2.122** $(\forall a)(\forall b)(((a \in \mathrm{N} \wedge b \in \mathrm{N}) \wedge a \cup \{a\} = b \cup \{b\}) \Rightarrow a = b)$

***Proof***

| | | |
|---|---|---|
| 1. | $(\forall a)(a \in \{a\})$ | T 2.6 |
| 2. | $(\forall a)(\forall b)(\forall c)(c \in b \Rightarrow c \in a \cup b)$ | T 2.30 |
| 3. | $(\forall a)(\forall b)(\forall c)(c \in a \cup b \Leftrightarrow (c \in a \vee c \in b))$ | T 2.18 |
| 4. | $(\forall a)(\forall b)(b \in \{a\} \Leftrightarrow b = a)$ | T 2.8 |
| 5. | $(\forall a)(a \in \mathrm{N} \Rightarrow a$ is transitive) | T 2.121 |
| 6. | $(\forall a)(a$ is transitive $\Leftrightarrow (\forall b)(\forall c)((b \in c \wedge c \in a) \Rightarrow b \in a))$ | Ax 2.31 |
| 7. | $(\forall a)(\forall b)(b \in \{a\} \Leftrightarrow a = b)$ | T 2.9 |
| 8. | $(\forall a)(\forall b)((\forall c)(c \in a \Leftrightarrow c \in b) \Rightarrow a = b)$ | Ax 2.1 |
| 9. | flag $d$ | FSUG |
| 10. | flag $e$ | FSUG |
| 11. | $(d \in \mathrm{N} \wedge e \in \mathrm{N}) \wedge d \cup \{d\} = e \cup \{e\}$ | ACP |
| 12. | $\neg(d = e)$ | AIP |
| 13. | $d \in \{d\}$ | UI 1 |
| 14. | $(\forall b)(\forall c)(c \in b \Rightarrow c \in d \cup b)$ | UI 2 |
| 15. | $(\forall c)(c \in \{d\} \Rightarrow c \in d \cup \{d\})$ | UI 14 |
| 16. | $d \in \{d\} \Rightarrow d \in d \cup \{d\}$ | UI 15 |
| 17. | $d \in d \cup \{d\}$ | MP 13, 16 |
| 18. | $d \cup \{d\} = e \cup \{e\}$ | S 11 |
| 19. | $d \in e \cup \{e\}$ | IS 17, 18 |
| 20. | $(\forall b)(\forall c)(c \in e \cup b \Leftrightarrow (c \in e \vee c \in b))$ | UI 3 |
| 21. | $(\forall c)(c \in e \cup \{e\} \Leftrightarrow (c \in e \vee c \in \{e\}))$ | UI 20 |
| 22. | $d \in e \cup \{e\} \Leftrightarrow (d \in e \vee d \in \{e\})$ | UI 21 |
| 23. | $d \in e \vee d \in \{e\}$ | BEl 19, 22 |
| 24. | $(\forall b)(b \in \{e\} \Leftrightarrow b = e)$ | UI 4 |
| 25. | $d \in \{e\} \Leftrightarrow d = e$ | UI 24 |
| 26. | $(d \in \{e\} \Rightarrow d = e) \wedge (d = e \Rightarrow d \in \{e\})$ | BE 25 |
| 27. | $d \in \{e\} \Rightarrow d = e$ | S 26 |
| 28. | $d \in e \vee d = e$ | D 23, 27 |
| 29. | $d \in e$ | DS 12, 28 |
| 30. | $d \in \mathrm{N} \wedge e \in \mathrm{N}$ | S 11 |

| | | |
|---|---|---|
| 31. | $e \in \mathrm{N}$ | S 30 |
| 32. | $e \in \mathrm{N} \Rightarrow e$ is transitive | UI 5 |
| 33. | $e$ is transitive | MP 31, 32 |
| 34. | flag $f$ | FSUG |
| 35. | $f \in d$ | ACP |
| 36. | $f \in d \wedge d \in e$ | C 29, 35 |
| 37. | $e$ is transitive $\Leftrightarrow$ $(\forall b)(\forall c)((b \in c \wedge c \in e) \Rightarrow b \in e)$ | UI 6 |
| 38. | $(\forall b)(\forall c)((b \in c \wedge c \in e) \Rightarrow b \in e)$ | BEI 33, 37 |
| 39. | $(\forall c)((f \in c \wedge c \in e) \Rightarrow f \in e)$ | UI 38 |
| 40. | $(f \in d \wedge d \in e) \Rightarrow f \in e$ | UI 39 |
| 41. | $f \in e$ | MP 36, 40 |
| 42. | $f \in d \Rightarrow f \in e$ | CP 35-41 |
| 43. | $(\forall c)(c \in d \Rightarrow c \in e)$ | UG 34-42 |
| 44. | $e \in \{e\}$ | UI 1 |
| 45. | $(\forall b)(\forall c)(c \in b \Rightarrow c \in e \cup b)$ | UI 2 |
| 46. | $(\forall c)(c \in \{e\} \Rightarrow c \in e \cup \{e\})$ | UI 45 |
| 47. | $e \in \{e\} \Rightarrow e \in e \cup \{e\}$ | UI 46 |
| 48. | $e \in e \cup \{e\}$ | MP 44, 47 |
| 49. | $e \in d \cup \{d\}$ | IS 18, 48 |
| 50. | $(\forall b)(\forall c)(c \in d \cup b \Leftrightarrow (c \in d \vee c \in b))$ | UI 3 |
| 51. | $(\forall c)(c \in d \cup \{d\} \Leftrightarrow (c \in d \vee c \in \{d\}))$ | UI 50 |
| 52. | $e \in d \cup \{d\} \Leftrightarrow (e \in d \vee e \in \{d\})$ | UI 51 |
| 53. | $e \in d \vee e \in \{d\}$ | BEI 49, 52 |
| 54. | $(\forall b)(b \in \{d\} \Leftrightarrow d = b)$ | UI 7 |
| 55. | $e \in \{d\} \Leftrightarrow d = e$ | UI 54 |
| 56. | $(e \in \{d\} \Rightarrow d = e) \wedge (d = e \Rightarrow e \in \{d\})$ | BE 55 |
| 57. | $e \in \{d\} \Rightarrow d = e$ | S 56 |
| 58. | $e \in d \vee d = e$ | D 53, 57 |
| 59. | $e \in d$ | DS 12, 58 |
| 60. | $d \in \mathrm{N}$ | S 30 |
| 61. | $d \in \mathrm{N} \Rightarrow d$ is transitive | UI 5 |
| 62. | $d$ is transitive | MP 60, 61 |
| 63. | flag $g$ | FSUG |
| 64. | $g \in e$ | ACP |
| 65. | $g \in e \wedge e \in d$ | C 59, 64 |
| 66. | $d$ is transitive $\Leftrightarrow$ $(\forall b)(\forall c)((b \in c \wedge c \in d) \Rightarrow b \in d)$ | UI 6 |
| 67. | $(\forall b)(\forall c)((b \in c \wedge c \in d) \Rightarrow b \in d)$ | BEI 62, 66 |
| 68. | $(\forall c)((g \in c \wedge c \in d) \Rightarrow g \in d)$ | UI 67 |
| 69. | $(g \in e \wedge e \in d) \Rightarrow g \in d$ | UI 68 |
| 70. | $g \in d$ | MP 65, 69 |

| | | |
|---|---|---|
| 71. | $g \in e \Rightarrow g \in d$ | CP 64-70 |
| 72. | $(\forall c)(c \in e \Rightarrow c \in d)$ | UG 63-71 |
| 73. | flag $h$ | FSUG |
| 74. | $h \in d \Rightarrow h \in e$ | UI 43 |
| 75. | $h \in e \Rightarrow h \in d$ | UI 72 |
| 76. | $(h \in d \Rightarrow h \in e) \wedge (h \in e \Rightarrow h \in d)$ | C 74, 75 |
| 77. | $h \in d \Leftrightarrow h \in e$ | BE 76 |
| 78. | $(\forall c)(c \in d \Leftrightarrow c \in e)$ | UG 73-77 |
| 79. | $(\forall b)((\forall c)(c \in d \Leftrightarrow c \in b) \Rightarrow d = b)$ | UI 8 |
| 80. | $(\forall c)(c \in d \Leftrightarrow c \in e) \Rightarrow d = e$ | UI 79 |
| 81. | $d = e$ | MP 78, 80 |
| 82. | $d = e \wedge \neg(d = e)$ | C 12, 81 |
| 83. | $\neg\neg(d = e)$ | IP 12-82 |
| 84. | $d = e$ | DN 83 |
| 85. | $((d \in \mathrm{N} \wedge e \in \mathrm{N}) \wedge d \cup \{d\} = e \cup \{e\}) \Rightarrow d = e$ | CP 11-84 |
| 86. | $(\forall b)(((d \in \mathrm{N} \wedge b \in \mathrm{N}) \wedge d \cup \{d\} = b \cup \{b\}) \Rightarrow d = b)$ | UG 10-85 |
| 87. | $(\forall a)(\forall b)(((a \in \mathrm{N} \wedge b \in \mathrm{N}) \wedge a \cup \{a\} = b \cup \{b\}) \Rightarrow a = b)$ | UG 9-86 |

□

## 2.14 Peano System

**Theorem 2.123** $(\exists! a)(\forall b)(b \in a \Leftrightarrow (b \in \mathrm{N} \times \mathrm{N} \wedge (\exists c)(b = (c, c \cup \{c\}))))$

***Proof***

| | | |
|---|---|---|
| 1. | $(\forall d)(\exists a)(\forall b)(b \in a \Leftrightarrow (b \in d \wedge (\exists c)(b = (c, c \cup \{c\}))))$ | AS 2.1 |
| 2. | $(\forall a)(\forall b)((\forall c)(c \in a \Leftrightarrow c \in b) \Rightarrow a = b)$ | Ax 2.1 |
| 3. | $(\exists a)(\forall b)(b \in a \Leftrightarrow (b \in \mathrm{N} \times \mathrm{N} \wedge (\exists c)(b = (c, c \cup \{c\}))))$ | UI 1 |
| 4. | flag $e$ | FSUG |
| 5. | flag $f$ | FSUG |
| 6. | $(\forall b)(b \in e \Leftrightarrow (b \in \mathrm{N} \times \mathrm{N} \wedge (\exists c)(b = (c, c \cup \{c\})))) \wedge$<br>$(\forall b)(b \in f \Leftrightarrow (b \in \mathrm{N} \times \mathrm{N} \wedge (\exists c)(b = (c, c \cup \{c\}))))$ | ACP |
| 7. | $(\forall b)(b \in e \Leftrightarrow (b \in \mathrm{N} \times \mathrm{N} \wedge (\exists c)(b = (c, c \cup \{c\}))))$ | S 6 |
| 8. | $(\forall b)(b \in f \Leftrightarrow (b \in \mathrm{N} \times \mathrm{N} \wedge (\exists c)(b = (c, c \cup \{c\}))))$ | S 6 |
| 9. | flag $g$ | FSUG |
| 10. | $g \in e \Leftrightarrow (g \in \mathrm{N} \times \mathrm{N} \wedge (\exists c)(g = (c, c \cup \{c\})))$ | UI 7 |
| 11. | $g \in f \Leftrightarrow (g \in \mathrm{N} \times \mathrm{N} \wedge (\exists c)(g = (c, c \cup \{c\})))$ | UI 8 |
| 12. | $g \in e \Leftrightarrow g \in f$ | BT 10, 11 |
| 13. | $(\forall c)(c \in e \Leftrightarrow c \in f)$ | UG 9-12 |
| 14. | $(\forall b)((\forall c)(c \in e \Leftrightarrow c \in b) \Rightarrow e = b)$ | UI 2 |
| 15. | $(\forall c)(c \in e \Leftrightarrow c \in f) \Rightarrow e = f$ | UI 14 |
| 16. | $e = f$ | MP 13, 15 |

| | | |
|---|---|---|
| 17. | $((\forall b)(b \in e \Leftrightarrow (b \in \mathrm{N} \times \mathrm{N} \wedge (\exists c)(b = (c, c \cup \{c\}))))\wedge$ $(\forall b)(b \in f \Leftrightarrow (b \in \mathrm{N} \times \mathrm{N} \wedge (\exists c)(b = (c, c \cup \{c\}))))) \Rightarrow$ $e = f$ | CP 6-16 |
| 18. | $(\forall h)(((\forall b)(b \in e \Leftrightarrow (b \in \mathrm{N} \times \mathrm{N} \wedge (\exists c)(b = (c, c \cup \{c\}))))\wedge$ $(\forall b)(b \in h \Leftrightarrow (b \in \mathrm{N} \times \mathrm{N} \wedge (\exists c)(b = (c, c \cup \{c\}))))) \Rightarrow$ $e = h)$ | UG 5-17 |
| 19. | $(\forall i)(\forall h)(((\forall b)(b \in i \Leftrightarrow (b \in \mathrm{N} \times \mathrm{N} \wedge (\exists c)(b = (c, c \cup \{c\}))))\wedge$ $(\forall b)(b \in h \Leftrightarrow (b \in \mathrm{N} \times \mathrm{N} \wedge (\exists c)(b = (c, c \cup \{c\}))))) \Rightarrow i = h)$ | UG 4-18 |
| 20. | $(\exists a)(\forall b)(b \in a \Leftrightarrow (b \in \mathrm{N} \times \mathrm{N} \wedge (\exists c)(b = (c, c \cup \{c\}))))\wedge$ $(\forall i)(\forall h)(((\forall b)(b \in i \Leftrightarrow (b \in \mathrm{N} \times \mathrm{N} \wedge (\exists c)(b = (c, c \cup \{c\}))))\wedge$ $(\forall b)(b \in h \Leftrightarrow (b \in \mathrm{N} \times \mathrm{N} \wedge (\exists c)(b = (c, c \cup \{c\}))))) \Rightarrow i = h)$ | C 3, 19 |
| 21. | $(\exists! a)(\forall b)(b \in a \Leftrightarrow (b \in \mathrm{N} \times \mathrm{N} \wedge (\exists c)(b = (c, c \cup \{c\}))))$ | UQ 20 |

□

**Axiom 2.32 (Definition of Constant "successor" from Theorem 2.123)**
$(\forall a)(a \in \text{successor} \Leftrightarrow (a \in \mathrm{N} \times \mathrm{N} \wedge (\exists b)(a = (b, b \cup \{b\}))))$

**Theorem 2.124** successor is a relation

***Proof***

| | | |
|---|---|---|
| 1. | $(\forall b)(b \in \text{successor} \Leftrightarrow (b \in \mathrm{N} \times \mathrm{N} \wedge (\exists c)(b = (c, c \cup \{c\}))))$ | Ax 2.32 |
| 2. | $(\forall a)(a \text{ is a relation} \Leftrightarrow (\forall b)(b \in a \Rightarrow (\exists c)(\exists d)(b = (c, d))))$ | Ax 2.22 |
| 3. | flag $e$ | FSUG |
| 4. | $e \in \text{successor}$ | ACP |
| 5. | $e \in \text{successor} \Leftrightarrow (e \in \mathrm{N} \times \mathrm{N} \wedge (\exists c)(e = (c, c \cup \{c\})))$ | UI 1 |
| 6. | $e \in \mathrm{N} \times \mathrm{N} \wedge (\exists c)(e = (c, c \cup \{c\}))$ | BEI 4, 5 |
| 7. | $(\exists c)(e = (c, c \cup \{c\}))$ | S 6 |
| 8. | $e = (f, f \cup \{f\})$ | EI 7, flag $f$ |
| 9. | $(\exists d)(e = (f, d))$ | EG 8 |
| 10. | $(\exists c)(\exists d)(e = (c, d))$ | EG 9 |
| 11. | $e \in \text{successor} \Rightarrow (\exists c)(\exists d)(e = (c, d))$ | CP 4-10 |
| 12. | $(\forall b)(b \in \text{successor} \Rightarrow (\exists c)(\exists d)(b = (c, d)))$ | UG 3-11 |
| 13. | successor is a relation $\Leftrightarrow$ $(\forall b)(b \in \text{successor} \Rightarrow (\exists c)(\exists d)(b = (c, d)))$ | UI 2 |
| 14. | successor is a relation | BEI 12, 13 |

□

**Theorem 2.125** successor is a function

***Proof***

| | | |
|---|---|---|
| 1. | $(\forall a)(a \in \text{successor} \Leftrightarrow (a \in \mathrm{N} \times \mathrm{N} \wedge (\exists b)(a = (b, b \cup \{b\}))))$ | Ax 2.32 |
| 2. | $(\forall a)(\forall b)(\forall c)(\forall d)((a, b) = (c, d) \Rightarrow (a = c \wedge b = d))$ | T 2.86 |
| 3. | successor is a relation | T 2.124 |
| 4. | $(\forall a)(a \text{ is a function} \Leftrightarrow (a \text{ is a relation} \wedge$ $(\forall b)(\forall c)(\forall d)(((b, c) \in a \wedge (b, d) \in a) \Rightarrow c = d)))$ | Ax 2.23 |
| 5. | flag $e$ | FSUG |
| 6. | flag $f$ | FSUG |
| 7. | flag $g$ | FSUG |
| 8. | $(e, f) \in \text{successor} \wedge (e, g) \in \text{successor}$ | ACP |
| 9. | $(e, f) \in \text{successor}$ | S 8 |
| 10. | $(e, f) \in \text{successor} \Leftrightarrow$ $((e, f) \in \mathrm{N} \times \mathrm{N} \wedge (\exists b)((e, f) = (b, b \cup \{b\})))$ | UI 1 |
| 11. | $(e, f) \in \mathrm{N} \times \mathrm{N} \wedge (\exists b)((e, f) = (b, b \cup \{b\}))$ | BEI 9, 10 |
| 12. | $(\exists b)((e, f) = (b, b \cup \{b\}))$ | S 11 |
| 13. | $(e, f) = (h, h \cup \{h\})$ | EI 12, flag $h$ |
| 14. | $(\forall b)(\forall c)(\forall d)((e, b) = (c, d) \Rightarrow (e = c \wedge b = d))$ | UI 2 |
| 15. | $(\forall c)(\forall d)((e, f) = (c, d) \Rightarrow (e = c \wedge f = d))$ | UI 14 |
| 16. | $(\forall d)((e, f) = (h, d) \Rightarrow (e = h \wedge f = d))$ | UI 15 |
| 17. | $(e, f) = (h, h \cup \{h\}) \Rightarrow (e = h \wedge f = h \cup \{h\})$ | UI 16 |
| 18. | $e = h \wedge f = h \cup \{h\}$ | MP 13, 17 |
| 19. | $e = h$ | S 18 |
| 20. | $f = h \cup \{h\}$ | S 18 |
| 21. | $f = e \cup \{e\}$ | IS 19, 20 |
| 22. | $(e, g) \in \text{successor}$ | S 8 |
| 23. | $(e, g) \in \text{successor} \Leftrightarrow$ $((e, g) \in \mathrm{N} \times \mathrm{N} \wedge (\exists b)((e, g) = (b, b \cup \{b\})))$ | UI 1 |
| 24. | $(e, g) \in \mathrm{N} \times \mathrm{N} \wedge (\exists b)((e, g) = (b, b \cup \{b\}))$ | BEI 22, 23 |
| 25. | $(\exists b)((e, g) = (b, b \cup \{b\}))$ | S 24 |
| 26. | $(e, g) = (i, i \cup \{i\})$ | EI 25, flag $i$ |
| 27. | $(\forall c)(\forall d)((e, g) = (c, d) \Rightarrow (e = c \wedge g = d))$ | UI 14 |
| 28. | $(\forall d)((e, g) = (i, d) \Rightarrow (e = i \wedge g = d))$ | UI 27 |
| 29. | $(e, g) = (i, i \cup \{i\}) \Rightarrow (e = i \wedge g = i \cup \{i\})$ | UI 28 |
| 30. | $e = i \wedge g = i \cup \{i\}$ | MP 26, 29 |
| 31. | $e = i$ | S 30 |
| 32. | $g = i \cup \{i\}$ | S 30 |
| 33. | $g = e \cup \{e\}$ | IS 31, 32 |
| 34. | $f = g$ | IT 21, 33 |
| 35. | $((e, f) \in \text{successor} \wedge (e, g) \in \text{successor}) \Rightarrow f = g$ | CP 8-34 |
| 36. | $(\forall d)(((e, f) \in \text{successor} \wedge (e, d) \in \text{successor}) \Rightarrow f = d)$ | UG 7-35 |
| 37. | $(\forall c)(\forall d)(((e, c) \in \text{successor} \wedge (e, d) \in \text{successor}) \Rightarrow c = d)$ | UG 6-36 |

| | | |
|---|---|---|
| 38. | $(\forall b)(\forall c)(\forall d)$ $(((b, c) \in \text{successor} \wedge (b, d) \in \text{successor}) \Rightarrow c = d)$ | UG 5-37 |
| 39. | successor is a relation $\wedge\ (\forall b)(\forall c)(\forall d)$ $(((b, c) \in \text{successor} \wedge (b, d) \in \text{successor}) \Rightarrow c = d)$ | C 3, 38 |
| 40. | successor is a function $\Leftrightarrow$ (successor is a relation $\wedge\ (\forall b)(\forall c)$ $(\forall d)(((b, c) \in \text{successor} \wedge (b, d) \in \text{successor}) \Rightarrow c = d))$ | UI 4 |
| 41. | successor is a function | BEl 39, 40 |

□

**Theorem 2.126** D(successor) $\subseteq$ N

***Proof***

| | | |
|---|---|---|
| 1. | $(\forall a)(\forall b)(\forall c)((a \text{ is a function} \wedge a \subseteq b \times c) \Rightarrow \mathrm{D}(a) \subseteq b)$ | T 2.106 |
| 2. | $(\forall a)(a \in \text{successor} \Leftrightarrow (a \in \mathrm{N} \times \mathrm{N} \wedge (\exists b)(a = (b, b \cup \{b\}))))$ | Ax 2.32 |
| 3. | $(\forall a)(\forall b)(a \subseteq b \Leftrightarrow (\forall c)(c \in a \Rightarrow c \in b))$ | Ax 2.9 |
| 4. | successor is a function | T 2.125 |
| 5. | $(\forall b)(\forall c)((\text{successor is a function} \wedge \text{successor} \subseteq b \times c) \Rightarrow$ $\mathrm{D}(\text{successor}) \subseteq b)$ | UI 1 |
| 6. | $(\forall c)((\text{successor is a function} \wedge \text{successor} \subseteq \mathrm{N} \times c) \Rightarrow$ $\mathrm{D}(\text{successor}) \subseteq \mathrm{N})$ | UI 5 |
| 7. | $(\text{successor is a function} \wedge \text{successor} \subseteq \mathrm{N} \times \mathrm{N}) \Rightarrow$ $\mathrm{D}(\text{successor}) \subseteq \mathrm{N}$ | UI 6 |
| 8. | flag $d$ | FSUG |
| 9. | $d \in$ successor | ACP |
| 10. | $d \in \text{successor} \Leftrightarrow (d \in \mathrm{N} \times \mathrm{N} \wedge (\exists b)(d = (b, b \cup \{b\})))$ | UI 2 |
| 11. | $d \in \mathrm{N} \times \mathrm{N} \wedge (\exists b)(d = (b, b \cup \{b\}))$ | BEl 9, 10 |
| 12. | $d \in \mathrm{N} \times \mathrm{N}$ | S 11 |
| 13. | $d \in \text{successor} \Rightarrow d \in \mathrm{N} \times \mathrm{N}$ | CP 9-12 |
| 14. | $(\forall c)(c \in \text{successor} \Rightarrow c \in \mathrm{N} \times \mathrm{N})$ | UG 8-13 |
| 15. | $(\forall b)(\text{successor} \subseteq b \Leftrightarrow (\forall c)(c \in \text{successor} \Rightarrow c \in b))$ | UI 3 |
| 16. | $\text{successor} \subseteq \mathrm{N} \times \mathrm{N} \Leftrightarrow (\forall c)(c \in \text{successor} \Rightarrow c \in \mathrm{N} \times \mathrm{N})$ | UI 15 |
| 17. | $\text{successor} \subseteq \mathrm{N} \times \mathrm{N}$ | BEl 14, 16 |
| 18. | successor is a function $\wedge$ successor $\subseteq \mathrm{N} \times \mathrm{N}$ | C 4, 17 |
| 19. | $\mathrm{D}(\text{successor}) \subseteq \mathrm{N}$ | MP 7, 18 |

□

**Theorem 2.127** $(\forall a)(a \in \mathrm{N} \Rightarrow (a, a \cup \{a\}) \in \text{successor})$

***Proof***

| | | |
|---|---|---|
| 1. | $(\forall a)(a \text{ is inductive} \Leftrightarrow (\phi \in a \wedge (\forall b)(b \in a \Rightarrow b \cup \{b\} \in a)))$ | Ax 2.29 |
| 2. | N is inductive | T 2.114 |
| 3. | $(\forall a)(\forall b)(\forall c)(\forall d)((c \in a \wedge d \in b) \Leftrightarrow (c, d) \in a \times b)$ | T 2.92 |
| 4. | $(\forall a)(a \in \text{successor} \Leftrightarrow (a \in \mathrm{N} \times \mathrm{N} \wedge (\exists b)(a = (b, b \cup \{b\}))))$ | Ax 2.32 |
| 5. | flag $e$ | FSUG |
| 6. | $e \in \mathrm{N}$ | ACP |
| 7. | $\mathrm{N} \text{ is inductive} \Leftrightarrow (\phi \in \mathrm{N} \wedge (\forall b)(b \in \mathrm{N} \Rightarrow b \cup \{b\} \in \mathrm{N}))$ | UI 1 |
| 8. | $\phi \in \mathrm{N} \wedge (\forall b)(b \in \mathrm{N} \Rightarrow b \cup \{b\} \in \mathrm{N})$ | BEl 2, 7 |
| 9. | $(\forall b)(b \in \mathrm{N} \Rightarrow b \cup \{b\} \in \mathrm{N})$ | S 8 |
| 10. | $e \in \mathrm{N} \Rightarrow e \cup \{e\} \in \mathrm{N}$ | UI 9 |
| 11. | $e \cup \{e\} \in \mathrm{N}$ | MP 6, 10 |
| 12. | $e \in \mathrm{N} \wedge e \cup \{e\} \in \mathrm{N}$ | C 6, 11 |
| 13. | $(\forall b)(\forall c)(\forall d)((c \in \mathrm{N} \wedge d \in b) \Leftrightarrow (c, d) \in \mathrm{N} \times b)$ | UI 3 |
| 14. | $(\forall c)(\forall d)((c \in \mathrm{N} \wedge d \in \mathrm{N}) \Leftrightarrow (c, d) \in \mathrm{N} \times \mathrm{N})$ | UI 13 |
| 15. | $(\forall d)((e \in \mathrm{N} \wedge d \in \mathrm{N}) \Leftrightarrow (e, d) \in \mathrm{N} \times \mathrm{N})$ | UI 14 |
| 16. | $(e \in \mathrm{N} \wedge e \cup \{e\} \in \mathrm{N}) \Leftrightarrow (e, e \cup \{e\}) \in \mathrm{N} \times \mathrm{N}$ | UI 15 |
| 17. | $(e, e \cup \{e\}) \in \mathrm{N} \times \mathrm{N}$ | BEl 12, 16 |
| 18. | $(e, e \cup \{e\}) = (e, e \cup \{e\})$ | IR |
| 19. | $(\exists b)((e, e \cup \{e\}) = (b, b \cup \{b\}))$ | EG 18 |
| 20. | $(e, e \cup \{e\}) \in \mathrm{N} \times \mathrm{N} \wedge (\exists b)((e, e \cup \{e\}) = (b, b \cup \{b\}))$ | C 17, 19 |
| 21. | $(e, e \cup \{e\}) \in \text{successor} \Leftrightarrow$ | |
| | $((e, e \cup \{e\}) \in \mathrm{N} \times \mathrm{N} \wedge (\exists b)((e, e \cup \{e\}) = (b, b \cup \{b\})))$ | UI 4 |
| 22. | $(e, e \cup \{e\}) \in \text{successor}$ | BEl 20, 21 |
| 23. | $e \in \mathrm{N} \Rightarrow (e, e \cup \{e\}) \in \text{successor}$ | CP 6-22 |
| 24. | $(\forall a)(a \in \mathrm{N} \Rightarrow (a, a \cup \{a\}) \in \text{successor})$ | UG 5-23 |

□

**Theorem 2.128** $\mathrm{N} \subseteq \mathrm{D}(\text{successor})$

***Proof***

| | | |
|---|---|---|
| 1. | $(\forall a)(a \in \mathrm{N} \Rightarrow (a, a \cup \{a\}) \in \text{successor})$ | T 2.127 |
| 2. | $(\forall a)(\forall b)(\forall c)((b, c) \in a \Rightarrow (b \in \cup\cup a \wedge c \in \cup\cup a))$ | T 2.99 |
| 3. | $(\forall a)(a \text{ is a function} \Rightarrow$ | |
| | $(\forall b)(b \in \mathrm{D}(a) \Leftrightarrow (b \in \cup\cup a \wedge (\exists c)((b, c) \in a))))$ | Ax 2.24 |
| 4. | successor is a function | T 2.125 |
| 5. | $(\forall a)(\forall b)(a \subseteq b \Leftrightarrow (\forall c)(c \in a \Rightarrow c \in b))$ | Ax 2.9 |
| 6. | flag $d$ | FSUG |
| 7. | $d \in \mathrm{N}$ | ACP |
| 8. | $d \in \mathrm{N} \Rightarrow (d, d \cup \{d\}) \in \text{successor}$ | UI 1 |

| | | |
|---|---|---|
| 9. | $(d, d \cup \{d\}) \in$ successor | MP 7, 8 |
| 10. | $(\forall b)(\forall c)((b, c) \in$ successor $\Rightarrow$ $(b \in \bigcup\bigcup$ successor $\wedge\, c \in \bigcup\bigcup$ successor$))$ | UI 2 |
| 11. | $(\forall c)((d, c) \in$ successor $\Rightarrow$ $(d \in \bigcup\bigcup$ successor $\wedge\, c \in \bigcup\bigcup$ successor$))$ | UI 10 |
| 12. | $(d, d \cup \{d\}) \in$ successor $\Rightarrow$ $(d \in \bigcup\bigcup$ successor $\wedge\, d \cup \{d\} \in \bigcup\bigcup$ successor$)$ | UI 11 |
| 13. | $d \in \bigcup\bigcup$ successor $\wedge\, d \cup \{d\} \in \bigcup\bigcup$ successor | MP 9, 12 |
| 14. | $d \in \bigcup\bigcup$ successor | S 13 |
| 15. | $(\exists c)((d, c) \in$ successor$)$ | EG 9 |
| 16. | $d \in \bigcup\bigcup$ successor $\wedge\, (\exists c)((d, c) \in$ successor$)$ | C 14, 15 |
| 17. | successor is a function $\Rightarrow (\forall b)(b \in$ D(successor) $\Leftrightarrow$ $(b \in \bigcup\bigcup$ successor $\wedge\, (\exists c)((b, c) \in$ successor$)))$ | UI 3 |
| 18. | $(\forall b)(b \in$ D(successor) $\Leftrightarrow$ $(b \in \bigcup\bigcup$ successor $\wedge\, (\exists c)((b, c) \in$ successor$)))$ | MP 4, 17 |
| 19. | $d \in$ D(successor) $\Leftrightarrow$ $(d \in \bigcup\bigcup$ successor $\wedge\, (\exists c)((d, c) \in$ successor$))$ | UI 18 |
| 20. | $d \in$ D(successor) | BEl 16, 19 |
| 21. | $d \in \mathrm{N} \Rightarrow d \in$ D(successor) | CP 7-20 |
| 22. | $(\forall c)(c \in \mathrm{N} \Rightarrow c \in$ D(successor)$)$ | UG 6-21 |
| 23. | $(\forall b)(\mathrm{N} \subseteq b \Leftrightarrow (\forall c)(c \in \mathrm{N} \Rightarrow c \in b))$ | UI 5 |
| 24. | $\mathrm{N} \subseteq$ D(successor) $\Leftrightarrow (\forall c)(c \in \mathrm{N} \Rightarrow c \in$ D(successor)$)$ | UI 23 |
| 25. | $\mathrm{N} \subseteq$ D(successor) | BEl 22, 24 |

□

**Theorem 2.129** D(successor) = N

***Proof***

| | | |
|---|---|---|
| 1. | D(successor) $\subseteq \mathrm{N}$ | T 2.126 |
| 2. | $\mathrm{N} \subseteq$ D(successor) | T 2.128 |
| 3. | $(\forall a)(\forall b)(a = b \Leftrightarrow (a \subseteq b \wedge b \subseteq a))$ | T 2.24 |
| 4. | D(successor) $\subseteq \mathrm{N} \wedge \mathrm{N} \subseteq$ D(successor) | C 1, 2 |
| 5. | $(\forall b)($D(successor) $= b \Leftrightarrow ($D(successor) $\subseteq b \wedge b \subseteq$ D(successor)$))$ | UI 3 |
| 6. | D(successor) $= \mathrm{N} \Leftrightarrow ($D(successor) $\subseteq \mathrm{N} \wedge \mathrm{N} \subseteq$ D(successor)$)$ | UI 5 |
| 7. | D(successor) $= \mathrm{N}$ | BEl 4, 6 |

□

**Theorem 2.130** R(successor) $\subseteq$ N

***Proof***

| | | |
|---|---|---|
| 1. | $(\forall a)(a \in \text{successor} \Leftrightarrow (a \in \text{N} \times \text{N} \wedge (\exists b)(a = (b, b \cup \{b\}))))$ | Ax 2.32 |
| 2. | $(\forall a)(\forall b)(a \subseteq b \Leftrightarrow (\forall c)(c \in a \Rightarrow c \in b))$ | Ax 2.9 |
| 3. | successor is a function | T 2.125 |
| 4. | $(\forall a)(\forall b)(\forall c)((a \text{ is a function} \wedge a \subseteq b \times c) \Rightarrow \text{R}(a) \subseteq c)$ | T 2.107 |
| 5. | flag $d$ | FSUG |
| 6. | $d \in \text{successor} \Leftrightarrow (d \in \text{N} \times \text{N} \wedge (\exists b)(d = (b, b \cup \{b\})))$ | UI 1 |
| 7. | $d \in \text{successor}$ | ACP |
| 8. | $d \in \text{N} \times \text{N} \wedge (\exists b)(d = (b, b \cup \{b\}))$ | BEl 6, 7 |
| 9. | $d \in \text{N} \times \text{N}$ | S 8 |
| 10. | $d \in \text{successor} \Rightarrow d \in \text{N} \times \text{N}$ | CP 7-9 |
| 11. | $(\forall c)(c \in \text{successor} \Rightarrow c \in \text{N} \times \text{N})$ | UG 5-10 |
| 12. | $(\forall b)(\text{successor} \subseteq b \Leftrightarrow (\forall c)(c \in \text{successor} \Rightarrow c \in b))$ | UI 2 |
| 13. | $\text{successor} \subseteq \text{N} \times \text{N} \Leftrightarrow (\forall c)(c \in \text{successor} \Rightarrow c \in \text{N} \times \text{N})$ | UI 12 |
| 14. | $\text{successor} \subseteq \text{N} \times \text{N}$ | BEl 11, 13 |
| 15. | successor is a function $\wedge$ successor $\subseteq \text{N} \times \text{N}$ | C 3, 14 |
| 16. | $(\forall b)(\forall c)((\text{successor is a function} \wedge \text{successor} \subseteq b \times c) \Rightarrow \text{R(successor)} \subseteq c)$ | UI 4 |
| 17. | $(\forall c)((\text{successor is a function} \wedge \text{successor} \subseteq \text{N} \times c) \Rightarrow \text{R(successor)} \subseteq c)$ | UI 16 |
| 18. | $(\text{successor is a function} \wedge \text{successor} \subseteq \text{N} \times \text{N}) \Rightarrow \text{R(successor)} \subseteq \text{N}$ | UI 17 |
| 19. | $\text{R(successor)} \subseteq \text{N}$ | MP 15, 18 |

□

**Theorem 2.131** $(\forall a)(a \in \text{N} \Rightarrow \text{successor}(a) = a \cup \{a\})$

***Proof***

| | | |
|---|---|---|
| 1. | $(\forall a)(a \in \text{N} \Rightarrow (a, a \cup \{a\}) \in \text{successor})$ | T 2.127 |
| 2. | successor is a function | T 2.125 |
| 3. | $(\forall a)(\forall b)(\forall c)((a \text{ is a function} \wedge (b, c) \in a) \Rightarrow b \in \text{D}(a))$ | T 2.104 |
| 4. | $(\forall a)(\forall b)((a \text{ is a function} \wedge b \in \text{D}(a)) \Rightarrow (b, a(b)) \in a)$ | Ax 2.26 |
| 5. | $(\forall a)(a \text{ is a function} \Leftrightarrow (a \text{ is a relation} \wedge (\forall b)(\forall c)(\forall d)(((b, c) \in a \wedge (b, d) \in a) \Rightarrow c = d)))$ | Ax 2.23 |
| 6. | flag $e$ | FSUG |
| 7. | $e \in \text{N}$ | ACP |
| 8. | $e \in \text{N} \Rightarrow (e, e \cup \{e\}) \in \text{successor}$ | UI 1 |
| 9. | $(e, e \cup \{e\}) \in \text{successor}$ | MP 7, 8 |
| 10. | successor is a function $\wedge$ $(e, e \cup \{e\}) \in \text{successor}$ | C 2, 9 |
| 11. | $(\forall b)(\forall c)((\text{successor is a function} \wedge (b, c) \in \text{successor}) \Rightarrow b \in \text{D(successor)})$ | UI 3 |

| | | |
|---|---|---|
| 12. | $(\forall c)((\text{successor is a function} \wedge (e, c) \in \text{successor}) \Rightarrow$ $e \in \text{D}(\text{successor}))$ | UI 11 |
| 13. | $(\text{successor is a function} \wedge (e, e \cup \{e\}) \in \text{successor}) \Rightarrow$ $e \in \text{D}(\text{successor})$ | UI 12 |
| 14. | $e \in \text{D}(\text{successor})$ | MP 10, 13 |
| 15. | $\text{successor is a function} \wedge e \in \text{D}(\text{successor})$ | C 2, 14 |
| 16. | $(\forall b)((\text{successor is a function} \wedge b \in \text{D}(\text{successor})) \Rightarrow$ $(b, \text{successor}(b)) \in \text{successor})$ | UI 4 |
| 17. | $(\text{successor is a function} \wedge e \in \text{D}(\text{successor})) \Rightarrow$ $(e, \text{successor}(e)) \in \text{successor}$ | UI 16 |
| 18. | $(e, \text{successor}(e)) \in \text{successor}$ | MP 15, 17 |
| 19. | $(e, \text{successor}(e)) \in \text{successor} \wedge (e, e \cup \{e\}) \in \text{successor}$ | C 9, 18 |
| 20. | $\text{successor is a function} \Leftrightarrow (\text{successor is a relation} \wedge$ $(\forall b)(\forall c)(\forall d)(((b, c) \in \text{successor} \wedge (b, d) \in \text{successor})$ $\Rightarrow c = d))$ | UI 5 |
| 21. | $\text{successor is a relation} \wedge (\forall b)(\forall c)(\forall d)$ $(((b, c) \in \text{successor} \wedge (b, d) \in \text{successor}) \Rightarrow c = d)$ | BEl 2, 20 |
| 22. | $(\forall b)(\forall c)(\forall d)(((b, c) \in \text{successor} \wedge (b, d) \in \text{successor})$ $\Rightarrow c = d)$ | S 21 |
| 23. | $(\forall c)(\forall d)(((e, c) \in \text{successor} \wedge (e, d) \in \text{successor})$ $\Rightarrow c = d)$ | UI 22 |
| 24. | $(\forall d)(((e, \text{successor}(e)) \in \text{successor} \wedge (e, d) \in \text{successor})$ $\Rightarrow \text{successor}(e) = d)$ | UI 23 |
| 25. | $((e, \text{successor}(e)) \in \text{successor} \wedge (e, e \cup \{e\}) \in \text{successor})$ $\Rightarrow \text{successor}(e) = e \cup \{e\}$ | UI 24 |
| 26. | $\text{successor}(e) = e \cup \{e\}$ | MP 19, 25 |
| 27. | $e \in \text{N} \Rightarrow \text{successor}(e) = e \cup \{e\}$ | CP 7-26 |
| 28. | $(\forall a)(a \in \text{N} \Rightarrow \text{successor}(a) = a \cup \{a\})$ | UG 6-27 |

□

**Theorem 2.132** $(\forall a)(a \in \text{N} \Rightarrow \text{successor}(a) \neq \phi)$

***Proof***

| | | |
|---|---|---|
| 1. | $(\forall a)(a \in \text{N} \Rightarrow \text{successor}(a) = a \cup \{a\})$ | T 2.131 |
| 2. | $(\forall a)(a \in \{a\})$ | T 2.6 |
| 3. | $(\forall a)(\forall b)(\forall c)(c \in b \Rightarrow c \in a \cup b)$ | T 2.30 |
| 4. | $(\forall a)((\exists b)(b \in a) \Rightarrow a \neq \phi)$ | T 2.48 |
| 5. | flag $d$ | FSUG |
| 6. | $d \in \text{N}$ | ACP |
| 7. | $d \in \text{N} \Rightarrow \text{successor}(d) = d \cup \{d\}$ | UI 1 |
| 8. | $\text{successor}(d) = d \cup \{d\}$ | MP 6, 7 |
| 9. | $d \in \{d\}$ | UI 2 |
| 10. | $(\forall b)(\forall c)(c \in b \Rightarrow c \in d \cup b)$ | UI 3 |

| | | |
|---|---|---|
| 11. | $(\forall c)(c \in \{d\} \Rightarrow c \in d \cup \{d\})$ | UI 10 |
| 12. | $d \in \{d\} \Rightarrow d \in d \cup \{d\}$ | UI 11 |
| 13. | $d \in d \cup \{d\}$ | MP 9, 12 |
| 14. | $(\exists b)(b \in d \cup \{d\})$ | EG 13 |
| 15. | $(\exists b)(b \in d \cup \{d\}) \Rightarrow d \cup \{d\} \neq \phi$ | UI 4 |
| 16. | $d \cup \{d\} \neq \phi$ | MP 14, 15 |
| 17. | $\text{successor}(d) \neq \phi$ | IS 8, 16 |
| 18. | $d \in \text{N} \Rightarrow \text{successor}(d) \neq \phi$ | CP 6-17 |
| 19. | $(\forall a)(a \in \text{N} \Rightarrow \text{successor}(a) \neq \phi)$ | UG 5-18 |

□

**Theorem 2.133** $(\forall a)(\forall b)(((a \in \text{N} \wedge b \in \text{N}) \wedge \text{successor}(a) = \text{successor}(b)) \Rightarrow a = b)$

***Proof***

| | | |
|---|---|---|
| 1. | $(\forall a)(a \in \text{N} \Rightarrow \text{successor}(a) = a \cup \{a\})$ | T 2.131 |
| 2. | $(\forall a)(\forall b)(((a \in \text{N} \wedge b \in \text{N}) \wedge a \cup \{a\} = b \cup \{b\}) \Rightarrow a = b)$ | T 2.122 |
| 3. | flag $c$ | FSUG |
| 4. | flag $d$ | FSUG |
| 5. | $(c \in \text{N} \wedge d \in \text{N}) \wedge \text{successor}(c) = \text{successor}(d)$ | ACP |
| 6. | $c \in \text{N} \wedge d \in \text{N}$ | S 5 |
| 7. | $\text{successor}(c) = \text{successor}(d)$ | S 5 |
| 8. | $c \in \text{N}$ | S 6 |
| 9. | $d \in \text{N}$ | S 6 |
| 10. | $c \in \text{N} \Rightarrow \text{successor}(c) = c \cup \{c\}$ | UI 1 |
| 11. | $\text{successor}(c) = c \cup \{c\}$ | MP 8, 10 |
| 12. | $d \in \text{N} \Rightarrow \text{successor}(d) = d \cup \{d\}$ | UI 1 |
| 13. | $\text{successor}(d) = d \cup \{d\}$ | MP 9, 12 |
| 14. | $c \cup \{c\} = \text{successor}(d)$ | IT 7, 11 |
| 15. | $c \cup \{c\} = d \cup \{d\}$ | IT 13, 14 |
| 16. | $(c \in \text{N} \wedge d \in \text{N}) \wedge c \cup \{c\} = d \cup \{d\}$ | C 6, 15 |
| 17. | $(\forall b)(((c \in \text{N} \wedge b \in \text{N}) \wedge c \cup \{c\} = b \cup \{b\}) \Rightarrow c = b)$ | UI 2 |
| 18. | $((c \in \text{N} \wedge d \in \text{N}) \wedge c \cup \{c\} = d \cup \{d\}) \Rightarrow c = d$ | UI 17 |
| 19. | $c = d$ | MP 16, 18 |
| 20. | $((c \in \text{N} \wedge d \in \text{N}) \wedge \text{successor}(c) = \text{successor}(d)) \Rightarrow c = d$ | CP 5-19 |
| 21. | $(\forall b)(((c \in \text{N} \wedge b \in \text{N}) \wedge \text{successor}(c) = \text{successor}(b))$ $\Rightarrow c = b)$ | UG 4-20 |
| 22. | $(\forall a)(\forall b)(((a \in \text{N} \wedge b \in \text{N}) \wedge \text{successor}(a) = \text{successor}(b))$ $\Rightarrow a = b)$ | UG 3-21 |

□

**Theorem 2.134** $(\forall a)(((a \subseteq \mathrm{N} \wedge \phi \in a) \wedge (\forall b)(b \in a \Rightarrow \text{successor}(b) \in a)) \Rightarrow a = \mathrm{N})$

***Proof***

| | | |
|---|---|---|
| 1. | $(\forall a)(\forall b)(a \subseteq b \Leftrightarrow (\forall c)(c \in a \Rightarrow c \in b))$ | Ax 2.9 |
| 2. | $(\forall a)(a \in \mathrm{N} \Rightarrow \text{successor}(a) = a \cup \{a\})$ | T 2.131 |
| 3. | $(\forall a)(((a \subseteq \mathrm{N} \wedge \phi \in a) \wedge (\forall b)(b \in a \Rightarrow b \cup \{b\} \in a)) \Rightarrow a = \mathrm{N})$ | T 2.118 |
| 4. | flag $d$ | FSUG |
| 5. | $(d \subseteq \mathrm{N} \wedge \phi \in d) \wedge (\forall b)(b \in d \Rightarrow \text{successor}(b) \in d)$ | ACP |
| 6. | $d \subseteq \mathrm{N} \wedge \phi \in d$ | S 5 |
| 7. | $(\forall b)(b \in d \Rightarrow \text{successor}(b) \in d)$ | S 5 |
| 8. | flag $e$ | FSUG |
| 9. | $e \in d \Rightarrow \text{successor}(e) \in d$ | UI 7 |
| 10. | $e \in d$ | ACP |
| 11. | $d \subseteq \mathrm{N}$ | S 6 |
| 12. | $(\forall b)(d \subseteq b \Leftrightarrow (\forall c)(c \in d \Rightarrow c \in b))$ | UI 1 |
| 13. | $d \subseteq \mathrm{N} \Leftrightarrow (\forall c)(c \in d \Rightarrow c \in \mathrm{N})$ | UI 12 |
| 14. | $(\forall c)(c \in d \Rightarrow c \in \mathrm{N})$ | BEI 11, 13 |
| 15. | $e \in d \Rightarrow e \in \mathrm{N}$ | UI 14 |
| 16. | $e \in \mathrm{N}$ | MP 10, 15 |
| 17. | $e \in \mathrm{N} \Rightarrow \text{successor}(e) = e \cup \{e\}$ | UI 2 |
| 18. | $\text{successor}(e) = e \cup \{e\}$ | MP 16, 17 |
| 19. | $\text{successor}(e) \in d$ | MP 9, 10 |
| 20. | $e \cup \{e\} \in d$ | IS 18, 19 |
| 21. | $e \in d \Rightarrow e \cup \{e\} \in d$ | CP 10-20 |
| 22. | $(\forall b)(b \in d \Rightarrow b \cup \{b\} \in d)$ | UG 8-21 |
| 23. | $(d \subseteq \mathrm{N} \wedge \phi \in d) \wedge (\forall b)(b \in d \Rightarrow b \cup \{b\} \in d)$ | C 6, 22 |
| 24. | $((d \subseteq \mathrm{N} \wedge \phi \in d) \wedge (\forall b)(b \in d \Rightarrow b \cup \{b\} \in d)) \Rightarrow d = \mathrm{N}$ | UI 3 |
| 25. | $d = \mathrm{N}$ | MP 23, 24 |
| 26. | $((d \subseteq \mathrm{N} \wedge \phi \in d) \wedge (\forall b)(b \in d \Rightarrow \text{successor}(b) \in d)) \Rightarrow d = \mathrm{N}$ | CP 5-25 |
| 27. | $(\forall a)(((a \subseteq \mathrm{N} \wedge \phi \in a) \wedge (\forall b)(b \in a \Rightarrow \text{successor}(b) \in a)) \Rightarrow a = \mathrm{N})$ | UG 4-26 |

□

**Theorem 2.135** successor : N $\to$ N

***Proof***

| | | |
|---|---|---|
| 1. | successor is a function | T 2.125 |
| 2. | D(successor) = N | T 2.129 |
| 3. | R(successor) $\subseteq$ N | T 2.130 |
| 4. | $(\forall a)(\forall b)(\forall c)$ <br> $(a : b \to c \Leftrightarrow ((a$ is a function $\wedge$ D$(a) = b) \wedge$ R$(a) \subseteq c))$ | Ax 2.27 |
| 5. | successor is a function $\wedge$ D(successor) = N | C 1, 2 |
| 6. | (successor is a function $\wedge$ D(successor) = N) $\wedge$ R(successor) $\subseteq$ N | C 3, 5 |
| 7. | $(\forall b)(\forall c)$(successor : $b \to c \Leftrightarrow$ <br> ((successor is a function $\wedge$ D(successor) $= b) \wedge$ R(successor) $\subseteq c$)) | UI 4 |
| 8. | $(\forall c)$(successor : N $\to c \Leftrightarrow$ <br> ((successor is a function $\wedge$ D(successor) = N) $\wedge$ R(successor) $\subseteq c$)) | UI 7 |
| 9. | successor : N $\to$ N $\Leftrightarrow$ <br> ((successor is a function $\wedge$ D(successor) = N) $\wedge$ R(successor) $\subseteq$ N) | UI 8 |
| 10. | successor : N $\to$ N | BEl 6, 9 |

□

**Axiom 2.33 (Definition of Predicate Which If Applied to the Arguments $\alpha$, $\beta$ and $\gamma$ Gives the Atomic Formula "$(\alpha, \beta, \gamma)$ is a Peano system")**
$(\forall a)(\forall b)(\forall c)((a, b, c)$ is a Peano system $\Leftrightarrow ((((c \in a \wedge b : a \to a) \wedge (\forall d)(d \in a \Rightarrow b(d) \neq c)) \wedge (\forall e)(\forall f)(((e \in a \wedge f \in a) \wedge b(e) = b(f)) \Rightarrow e = f)) \wedge (\forall g)(((g \subseteq a \wedge c \in g) \wedge (\forall h)(h \in g \Rightarrow b(h) \in g)) \Rightarrow g = a)))$

**Theorem 2.136** (N, successor, $\phi$) is a Peano system

***Proof***

| | | |
|---|---|---|
| 1. | $\phi \in$ N | T 2.115 |
| 2. | successor : N $\to$ N | T 2.135 |
| 3. | $(\forall d)(d \in$ N $\Rightarrow$ successor$(d) \neq \phi)$ | T 2.132 |
| 4. | $(\forall e)(\forall f)(((e \in$ N $\wedge f \in$ N$) \wedge$ successor$(e) =$ successor$(f))$ <br> $\Rightarrow e = f)$ | T 2.133 |
| 5. | $(\forall g)(((g \subseteq$ N $\wedge \phi \in g) \wedge (\forall h)(h \in g \Rightarrow$ successor$(h) \in g))$ <br> $\Rightarrow g =$ N$)$ | T 2.134 |
| 6. | $(\forall a)(\forall b)(\forall c)((a, b, c)$ is a Peano system $\Leftrightarrow$ <br> $((((c \in a \wedge b : a \to a) \wedge (\forall d)(d \in a \Rightarrow b(d) \neq c))\wedge$ <br> $(\forall e)(\forall f)(((e \in a \wedge f \in a) \wedge b(e) = b(f)) \Rightarrow e = f))\wedge$ <br> $(\forall g)(((g \subseteq a \wedge c \in g) \wedge (\forall h)(h \in g \Rightarrow b(h) \in g)) \Rightarrow g = a)))$ | Ax 2.33 |
| 7. | $\phi \in$ N $\wedge$ successor : N $\to$ N | C 1, 2 |

| | | |
|---|---|---|
| 8. | $(\phi \in \mathrm{N} \wedge \text{successor} : \mathrm{N} \to \mathrm{N}) \wedge$ $(\forall d)(d \in \mathrm{N} \Rightarrow \text{successor}(d) \neq \phi)$ | C 3, 7 |
| 9. | $((\phi \in \mathrm{N} \wedge \text{successor} : \mathrm{N} \to \mathrm{N}) \wedge$ $(\forall d)(d \in \mathrm{N} \Rightarrow \text{successor}(d) \neq \phi)) \wedge (\forall e)(\forall f)$ $(((e \in \mathrm{N} \wedge f \in \mathrm{N}) \wedge \text{successor}(e) = \text{successor}(f)) \Rightarrow e = f)$ | C 4, 8 |
| 10. | $(((\phi \in \mathrm{N} \wedge \text{successor} : \mathrm{N} \to \mathrm{N}) \wedge$ $(\forall d)(d \in \mathrm{N} \Rightarrow \text{successor}(d) \neq \phi)) \wedge (\forall e)(\forall f)$ $(((e \in \mathrm{N} \wedge f \in \mathrm{N}) \wedge \text{successor}(e) = \text{successor}(f)) \Rightarrow e = f)) \wedge$ $(\forall g)(((g \subseteq \mathrm{N} \wedge \phi \in g) \wedge$ $(\forall h)(h \in g \Rightarrow \text{successor}(h) \in g)) \Rightarrow g = \mathrm{N})$ | C 5, 9 |
| 11. | $(\forall b)(\forall c)((\mathrm{N}, b, c)$ is a Peano system $\Leftrightarrow$ $((((c \in \mathrm{N} \wedge b : \mathrm{N} \to \mathrm{N}) \wedge (\forall d)(d \in \mathrm{N} \Rightarrow b(d) \neq c)) \wedge$ $(\forall e)(\forall f)(((e \in \mathrm{N} \wedge f \in \mathrm{N}) \wedge b(e) = b(f)) \Rightarrow e = f)) \wedge$ $(\forall g)(((g \subseteq \mathrm{N} \wedge c \in g) \wedge (\forall h)(h \in g \Rightarrow b(h) \in g)) \Rightarrow g = \mathrm{N})))$ | UI 6 |
| 12. | $(\forall c)((\mathrm{N}, \text{successor}, c)$ is a Peano system $\Leftrightarrow$ $((((c \in \mathrm{N} \wedge \text{successor} : \mathrm{N} \to \mathrm{N}) \wedge$ $(\forall d)(d \in \mathrm{N} \Rightarrow \text{successor}(d) \neq c)) \wedge (\forall e)(\forall f)$ $(((e \in \mathrm{N} \wedge f \in \mathrm{N}) \wedge \text{successor}(e) = \text{successor}(f)) \Rightarrow e = f)) \wedge$ $(\forall g)(((g \subseteq \mathrm{N} \wedge c \in g) \wedge (\forall h)(h \in g \Rightarrow \text{successor}(h) \in g)) \Rightarrow$ $g = \mathrm{N})))$ | UI 11 |
| 13. | $(\mathrm{N}, \text{successor}, \phi)$ is a Peano system $\Leftrightarrow$ $((((\phi \in \mathrm{N} \wedge \text{successor} : \mathrm{N} \to \mathrm{N}) \wedge$ $(\forall d)(d \in \mathrm{N} \Rightarrow \text{successor}(d) \neq \phi)) \wedge (\forall e)(\forall f)$ $(((e \in \mathrm{N} \wedge f \in \mathrm{N}) \wedge \text{successor}(e) = \text{successor}(f)) \Rightarrow e = f)) \wedge$ $(\forall g)(((g \subseteq \mathrm{N} \wedge \phi \in g) \wedge (\forall h)(h \in g \Rightarrow \text{successor}(h) \in g))$ $\Rightarrow g = \mathrm{N}))$ | UI 12 |
| 14. | $(\mathrm{N}, \text{successor}, \phi)$ is a Peano system | BEl 10, 13 |

□

## References


1. Mendelson, E.: Number Systems and the Foundations of Analysis. Dover, New York (2008)
2. Jech, T.: Set Theory. Springer, New York (2002)
3. Mendelson, E.: Introduction to Mathematical Logic, 3rd edn. Wadsworth, Pacific Grove (1987)

FSC
www.fsc.org
MIX
Papier | Fördert gute Waldnutzung
FSC® C083411

Zeitfracht Medien GmbH
Ferdinand-Jühlke-Straße 7
99095 Erfurt, Deutschland
produktsicherheit@kolibri360.de

# BIN TRAVERLER FORM

Cut By: Jose Franco Qty 31 Date 09-21-26

Scanned By:________________ Qty______ Date__________

Scanned Batch ID's

Notes / Exceptions